STERLING
New York

An Imprint of Sterling Publishing
387 Park Avenue South
New York, NY 10016

© 2013 by Quintet Publishing Ltd

ISBN 978-1-4549-1190-6

This book was designed, conceived and produced by
Quintet Publishing Ltd
4th Floor, Sheridan House
112/116A Western Road
Hove, East Sussex
BN3 1DD

Design and editorial content: Tall Tree Ltd
Art Director: Michael Charles
Editorial Manager: Emma Bastow
Publisher: Mark Searle
QTT.SSY

Distributed in Canada by Sterling Publishing
c/o Canadian Manda Group, 165 Dufferin Street
Toronto, Ontario, Canada M6K 3H6

For information about custom editions, special sales, and premium and corporate purchases, please contact Sterling Special Sales at 800-805-5489 or specialsales@sterlingpublishing.com.

Manufactured in China

2 4 6 8 10 9 7 5 3 1

www.sterlingpublishing.com

SOLAR SYSTEM

DAVID W. HUGHES AND CAROLE STOTT

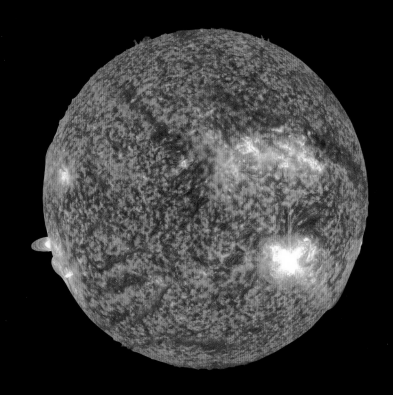

STERLING
New York

CONTENTS

▶ **Curiosity**

Curiosity is here pictured in the Gale Crater on October 31, 2012, where its first rock sample collection took place.

INTRODUCTION

The Solar System comprises our local star, the Sun, together with a large number of objects that orbit around it. These include the eight planets, dwarf planets, nearly 150 moons, and countless asteroids and comets. Modern cosmogonists (scientists who study the origin of the Universe) believe that the Solar System formed about 4.6 billion years ago from a vast cloud of gas and dust called the solar nebula. Over millions of years this spinning cloud collapsed under the force of gravity and flattened into a pancake-shaped disk that began rotating faster and faster as it contracted. The Sun formed out of the dense central part of the disk. Near the hot, central region, only rocky particles and metals could withstand the heat and remain in solid form. All other materials vaporized. Over time, these rock and metal particles clumped together to form planetesimals (small bodies of rock). In the cooler, outer regions of the disk a similar process occurred, but there the materials that clumped together comprised water ice, ammonia, rock, and methane ices.

These planetesimals formed over tens of millions of years, but the stage in which they finally became planets may have happened quickly, about 4.6 billion years ago. Once the planetesimals had reached a certain size, their gravity was strong enough to draw in more and more material in a rapid process. Planetesimals came together to form protoplanets, which then underwent a series of massive collisions with each other to form the inner rocky planets Mercury, Venus, Earth, and Mars, and the cores of the outer planets Jupiter, Saturn, Uranus, and Neptune. The outer planets, containing rock and ice, had enough mass to attract vast amounts of gases, which surrounded the planetary cores to become dense atmospheres.

▶ **The Solar System forms**

This artist's impression shows the vast cloud of gas and dust known as the solar nebula as a disk that is spinning ever faster. Instabilities in the disk caused regions within it to condense into rings under the influence of gravity. Gradually, small planetesimals of rock or rock and ice formed in the rings through the accretion of matter, and continued to grow larger until they eventually formed protoplanets. Many of the leftover planetesimals probably became comets and asteroids.

Mercury (p32)

The planet nearest the Sun has the fastest orbit in the Solar System, taking just 88 Earth days to circle the Sun at an average distance of 36 million miles (57.9 million km).

Earth (p48)

The only planet in a zone temperate enough for liquid water to exist, Earth orbits the Sun in 365.65 days at an average distance of 92.9 million miles (149.6 million km).

Jupiter (p88)

Jupiter, the largest planet in the Solar System, has such a strong gravitational field that it hoovers up much of the debris flying around the Solar System. It orbits the Sun in 11.86 Earth years at an average distance of 483.4 million miles (778.4 million km).

Mars (p72)

Known as the red planet because of the iron oxide rock and dust that covers its surface, Mars orbits the Sun in 697 Earth days at an average distance of 141.6 million miles (227.9 million km).

Venus (p40)

Sometimes known as "Earth's twin" because of its similar size to Earth, Venus orbits the Sun in 224.7 Earth days at an average distance of 67.2 million miles (108.2 million km).

◀ The Sun (p22)

The Sun contains 750 times the mass of all the Solar System's planets put together. In its core, nuclear reactions turn hydrogen into helium to generate huge amounts of energy. The energy is carried toward the surface where it escapes as radiation.

Family of planets

The Solar System is 9,300 billion miles (15 billion km) across. The inner region contains the Sun and the four rocky planets. Beyond these lies a belt of asteroids known as the Main Belt, followed by the four gas giant planets. Farther out is a vast region filled with comets. The planets occupy a zone extending just 3.25 billion miles (6 billion km) from the Sun. Most objects in the Solar System have orbits in the shape of ellipses (stretched circles), and all the planets and almost all the asteroids orbit the Sun in the same direction.

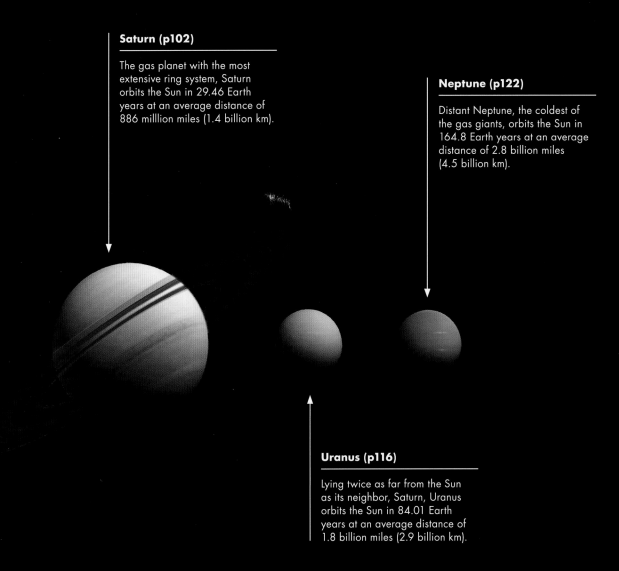

Saturn (p102)

The gas planet with the most extensive ring system, Saturn orbits the Sun in 29.46 Earth years at an average distance of 886 milllion miles (1.4 billion km).

Neptune (p122)

Distant Neptune, the coldest of the gas giants, orbits the Sun in 164.8 Earth years at an average distance of 2.8 billion miles (4.5 billion km).

Uranus (p116)

Lying twice as far from the Sun as its neighbor, Saturn, Uranus orbits the Sun in 84.01 Earth years at an average distance of 1.8 billion miles (2.9 billion km).

The rocky planets

The rocky planets have few or no moons and no rings. Early in their history the materials from which they are made separated into a metallic core and a rocky mantle and crust. Each of the four is quite different. Venus and Mars have atmospheres of carbon dioxide, but Venus's is extremely thick, while that of Mars is very thin. Mercury has almost no atmosphere at all, and Earth's is rich in oxygen and nitrogen. Earth also has a very large Moon—the largest in the Solar System proportionate to its planet's size.

The gas giants

The four gas giants have much in common. All have cores composed of rock and ice, surrounded by a liquid mantle of hydrogen and helium, or ammonia, methane, and water ices. Each has a turbulent atmosphere and a strong magnetic field, of which Jupiter's is the strongest, at least 20,000 times as strong as that of Earth. Each gas giant has a large number of moons, and ring systems made of rock, dust, and ice. In 2006, Pluto was reclassified as a dwarf planet, reducing the total number of planets in the Solar System to eight.

DISCOVERING THE SOLAR SYSTEM

Early astronomy was dominated by questions about the nature of the Solar System. To the ancient astronomers, the stars were pinpoints of light forming a celestial backdrop against which planetary movements could be plotted. But astronomers in those distant days did not know what other planets were like, the form of the planetary orbits, or the mechanism that kept planets moving. The search for answers stimulated the scientific enquiry of the Renaissance.

By the 19th century, astronomers were concentrating on the physics of the stars. They tried to understand where the Universe came from. Larger and larger telescopes were built, but they were mainly used to observe faint stars and nebulae.

It was the "Space Age" of the 20th century that revolutionized planetary studies. Instruments can now be sent from Earth to nearby worlds. Other planets have been found to have ice caps and dried-up rivers. Planetary surfaces are patterned with mountains and planes, impact craters and volcanoes. As space exploration has become more sophisticated, the few images from a fleeting fly-by have been replaced by huge data sets obtained by semi-permanent orbiting spacecraft and landers.

The last decade has seen the first fly-by of Pluto, the first comet nucleus orbit mission, and a return to Mercury and Mars. Even so, many of the large satellites of the gas giant planets cry out for detailed investigation. There is still much to explore.

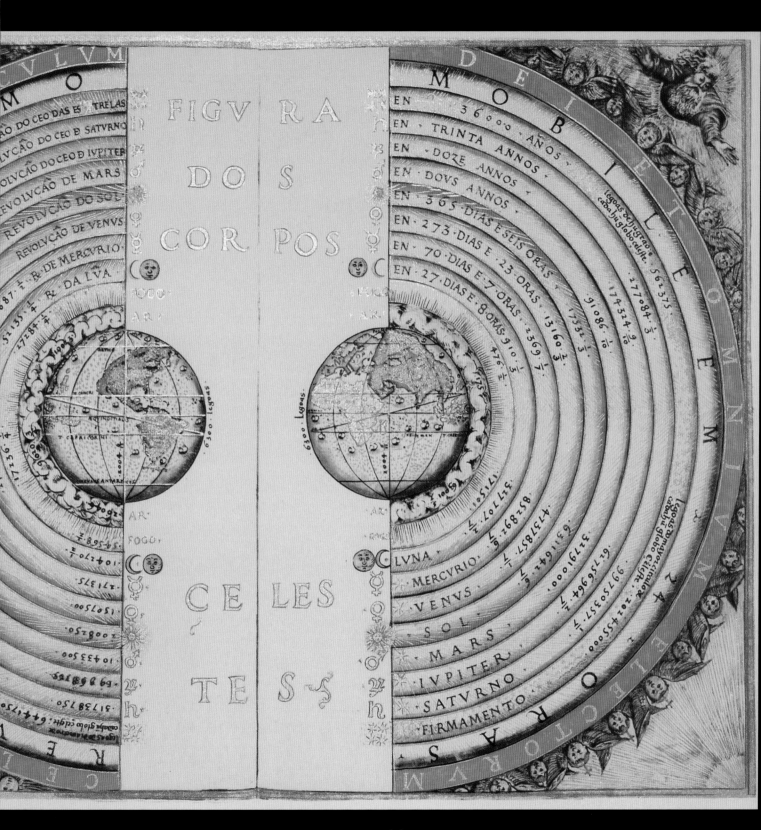

▲ **Center of the cosmos**

This 1568 illustration of the Solar System by Portuguese cosmographer Bartolomeu Velho places Earth at the center of the system. This geocentric model provides a conception of a cosmos with Earth at the orbital center of all celestial bodies. It was the predominant cosmological system in many ancient civilizations.

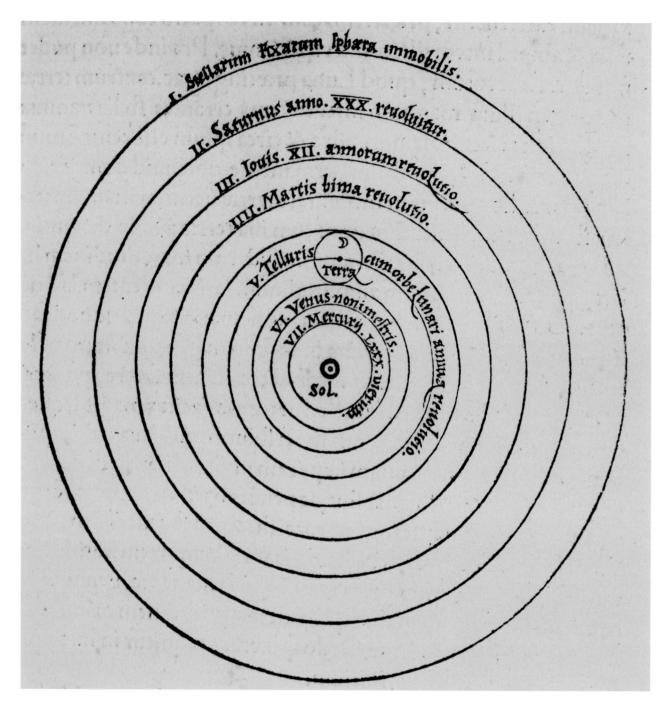

▲ **Heliocentric model**

This illustration shows the heliocentric system proposed by Nicolaus Copernicus in his book *On the Revolutions of the Celestial Spheres*, published in 1543. Copernicus's original manuscript for the book still survives, and this drawing of his model was made by Copernicus himself. Like Ptolemy before him, Copernicus conceived a Solar System in which the planets orbit in perfect circles. Some decades after Copernicus's death, German astronomer Johannes Kepler was the first to realize that the orbits were in fact elliptical.

The astronomy revolution

In 1608, the German-born Dutch spectacle maker Hans Lipperhey, widely credited with inventing the telescope, applied for a patent to protect his invention. Although the patent was declined, because his device was considered too simple, his application sparked interest across Europe.

By 1609, the Italian mathematician and scientist Galileo Galilei had gathered enough information about the telescope to build his own. His first was one of three times magnifying power, but he continued improving the design and by October 1609, had a telescope of 20 times magnification.

This was the instrument that revolutionized astronomy. Galileo turned his telescope toward the Sun, the Moon, Venus, and Jupiter, and almost immediately made three sensational discoveries.

For the first time, Galileo had evidence that the Universe was not as the Catholic Church claimed it to be. In the early 17th century, the Church was defending the traditional view of the cosmos based on the writings of the ancient Greek philosopher Aristotle, in whose model the Sun, stars, and planets all revolved around Earth. Galileo saw that the Moon was not a glowing heavenly object, but another world entirely, with mountains and distinct landscape features. Venus, which was thought to be a star, was clearly a disk. Furthermore it changed shape in the same way the Moon changed shape as it went through its monthly cycle, going from crescent to half disk to full disk. Jupiter was also a disk, but it was accompanied by four tiny points of light.

Galileo soon made sense of these strange observations. Venus could only change shape in this way if it was a spherical body going around the Sun. The shape changed because we were seeing portions of the planet in sunlight or shade as it orbited. The bright spots accompanying Jupiter were its moons. They changed position from night to night as they passed along their orbits around the planet.

Heliocentric Universe

Galileo had proved that the accepted view of the Universe was wrong. Venus orbited the Sun, and it was obvious that Jupiter's moons orbited Jupiter, not Earth. The Sun at the center of the Solar System, however, was not originally Galileo's idea, but that of a Polish astronomer called Nicolaus Copernicus who, in a book published just before his death in 1543, had provided an explanation for the movements of the planets by placing the Sun at the center of the system. Galileo's discoveries provided hard evidence to support Copernicus's model of a heliocentric Universe.

Elliptical orbits

The time it took planets to move between certain positions on their orbits enabled astronomers to calculate the ratios of planetary orbital sizes. It was realized that Jupiter was five times farther away from

the Sun than Earth, and Saturn was nine and a half times farther away. But why was the Moon going around Earth, where everything else was going around the Sun? Two things were not known. The first was the exact shape of the orbit; the second was the mechanism that was responsible for moving the planets. The first of these problems was solved by the German astronomer Johannes Kepler. In 1609, he found that the planetary orbits were elliptical and not circular. The Sun was at one of the foci of each ellipse.

◀ Copernicus

Nicolaus Copernicus's heliocentric model of the Solar System disagreed with the prevailing view in his time that Earth was the stationary center of the Universe. In the second century CE, Ptolemy had worked out a system that employed epicycles to account for observations that contradicted a simple Earth-centric model of the Universe. Copernicus's new model was mathematically far more elegant, but supporting evidence for his theory was not provided until Galileo made his breakthrough observations of the phases of Venus.

◀ Galileo

Acknowledged as one of the fathers of modern science, Galileo Galilei made his breakthrough observations using a telescope he made himself. Flaws in the lens design and the poor quality of the glass meant that the images were blurred and distorted, but it was still good enough for Galileo to view the phases of Venus, see craters on the Moon, and pick out Jupiter's four main moons.

was near perihelion, its closest point to the Sun, than it was when at the aphelion of the orbit, the most distant point. The third thing that Kepler discovered, in 1619, was that the square of the orbital period was proportional to the cube of the average distance between the planet and the Sun. However, Kepler had no idea what pushed the planets along their orbits, or what stopped them from falling into the Sun. The physics of planetary motions appeared only around 1687, when the English physicist Isaac Newton introduced the concept of universal gravity.

The force of gravity

One of the greatest science books ever published was Isaac Newton's *Principia*, which came out in 1687. Newton realized also varies as a function of the inverse square of the distance between the planet and the Sun. Newton suggested that the force was universal, tying the whole cosmos together. The force that made an apple fall to the ground was of the same nature as the force that kept the Moon orbiting Earth, Earth orbiting the Sun, and stars like the Sun orbiting the center of the galaxy.

This new theory allowed the calculation of the masses of orbiting bodies. Their distances and durations of orbit were all that was needed. The fact that Earth orbited the Sun, and the Moon orbited Earth, meant that the mass of the Sun and Earth could be calculated. The Sun turned out to be 335,000 times more massive than Earth.

◄ William Herschel

The German-born British astronomer Herschel discovered Uranus and two of its moons in 1781 while searching for double stars. He initially thought Uranus was a comet. Herschel built some of the finest telescopes of his day, making more than 400 of them during his lifetime, the largest a reflecting telescope with a primary mirror 49.5 in (1.26 m) in diameter.

▶ 2012 transit of Venus

This image from NASA's Solar Dynamics Observatory shows Venus as it nears the disk of the Sun on June 5, 2012. Venus's 2012 transit was the last such event until 2117. Observations of the transit of Venus in 1639 allowed English astronomer Jeremiah Horrocks to make one of the earliest calculations of the distance of Earth from the Sun.

Viewing the planets

With the advent of the astronomical telescope, the planetary "point sources" that can be seen by the eye were converted into disks. Knowing the distances to the planets, the sizes of the disks could be converted into actual diameters and volumes. The orbital sizes and periods of their newly discovered moons indicated that Jupiter and Saturn were 318 and 95 times more massive than Earth. Knowing the volumes meant that the planetary densities could be calculated. For example, astronomers quickly realized that Earth was 5.52 times as dense as water, Mars was 3.95 times as dense, and Jupiter and Saturn respectively 1.34 and 0.70 times as dense. These values could be compared with the densities of iron, rock, ice, and the air we breathe. Here the multipliers are 7.9, 2.9, 0.98, and 0.0013 times the density of water. So, knowing the mean density of a planet provided a strong clue to its composition.

A new planet was discovered completely by accident by William Herschel in 1781. This caused a sensation. Finding Uranus doubled the size of the Solar System. And things did not end there. The orbit of Uranus was being pushed and pulled by an unknown planet even farther away from the Sun. Detailed calculations about where this planet might be led to the discovery of Neptune in 1846.

A stable system

In the 18th century, French astronomer Pierre Laplace proved that the Solar System was stable. He found that the planetary orbits changed very little with time, and that the ordering of the planets today is the same as it was at the beginning. The nearest a passing star has come to the Sun is about 35 times farther away than Neptune. The Solar System has never been stirred up.

Perhaps surprisingly, by the beginning of the 20th century important questions remained unresolved. By then, it was clear that the Solar System was just one small part of a larger galaxy, but most astronomers thought the galaxy was the whole Universe.

Many galaxies

Evidence that the Universe was far larger than previously imagined came from information gathered by the Hooker Telescope at the Mount Wilson Observatory in California, a 100-in (2.5-m) telescope that was the largest in the world from 1917 to 1948. The Hooker used a large mirror to collect as much light as possible, seeing far fainter objects in the sky. It revealed the spiral structure of the cloudy nebulae. It was fast becoming clear that not only was the Sun not the center of the Milky Way galaxy, but that the galaxy itself was but one of many, and that each nebula was itself another distant galaxy.

Expanding Universe

Measuring the distance of stars and galaxies remained a tricky problem. The farther away a star is, the fainter it appears, but not all stars shine with the same brightness, so this on its own is not enough to calculate distance. Help came in the shape of a particular kind of variable star, the Cepheid variable. These stars change brightness at regular intervals, and the rate at which they change is related to their brightness. Find a Cepheid variable in a galaxy, and you can work out how far away that galaxy is.

One of the astronomers observing from Mount Wilson, Edwin Hubble, discovered a Cepheid variable in the Andromeda Nebula. He calculated that it was 2.5 million light

years away. Now renamed the Andromeda Galaxy, it was clearly far away from the Milky Way, but Andromeda is also the closest major galaxy to ours. It was evident, therefore, that the Universe was far larger than most had suspected.

Not only was the Universe far larger, but it was expanding. Hubble discovered this by measuring the change in wavelength of the light reaching us from other galaxies. A phenomenon called the Doppler effect means that the wavelength of light reaching us from an object that is moving away becomes longer, meaning that visible light shifts toward red in the spectrum. If the object is moving toward us, the wavelength becomes shorter, shifting toward blue in the spectrum. Hubble measured the red-shift and blue-shift of various galaxies. He found that Andromeda was blue-shifting—it is moving toward us, and may collide with the Milky Way one day. But other galaxies are red-shifting. And not only that, but the farther away they are, the more they are red-shifting. This means that the Universe is expanding, and it was later shown that this expansion is itself accelerating. But it does not mean that we are at the center of the Universe. From whichever galaxy you are looking, other galaxies are accelerating away.

Big Bang

Knowing that the Universe is expanding, we can extrapolate backward in time to the moment when it was just a single point. In 1927, Belgian mathematician George-Henri Lemaître did this, describing his idea as the "hypothesis of the primeval atom"—later disparagingly dubbed the "Big Bang" by astronomer Fred Hoyle, who favored a "steady state" theory. Confirmation of the Big Bang came in the form of the discovery of cosmic microwave background radiation, which pervades space and is explained as a relic from the early Universe. We now know that the Universe is 13.8 billion years old.

▲ Luna 2

Two years after Sputnik 1, the Soviet Union put the first artificial object on the Moon. Luna 2 hit the surface of the Moon east of Mare Serenitatis on September 14, 1959. It stopped transmitting when it landed. The spacecraft carried a magnetometer and a geiger counter. It confirmed that the Moon had no magnetic field and was not surrounded by radiation belts.

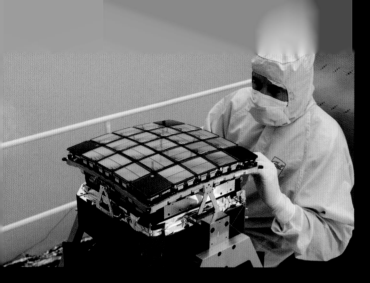

Other planetary systems

The Universe contains hundreds of billions of galaxies, each containing on average about a trillion stars. How many of these stars have planetary systems? And just how typical is the planetary system that we live in? When it comes to the first of these questions, we must consider metallicity, spin, and mass. The first generation of stars—those produced just after the Big Bang and known by astronomers as Population III stars—contain very few metals, and planets cannot be formed if there is no earthy and icy material. But stars that were formed after some of the more massive Population III stars had exploded as supernovae and seeded galactic clouds with metallic atoms, could have planets. These later-generation stars, such as the Sun make up half the stars in the sky.

Fragment, spin, and mass

Only about a third of the star-forming fragments have spins that are favorable to planet formation, and half of these form close binaries and not single stars with planets. Finally, there is the question of mass. The mass of the cloud is proportional to the mass of the central star, and the time it takes to form a planet is a function of the density. As the Universe is only about 14 billion years old, not enough time has elapsed for planets to have formed around stars that are less massive than about an eighth of the mass of the Sun. This rules out half the stars that have the right spin speed and metallicity. It follows that only about 4 percent (one in 25) of all stars in the Universe have planets.

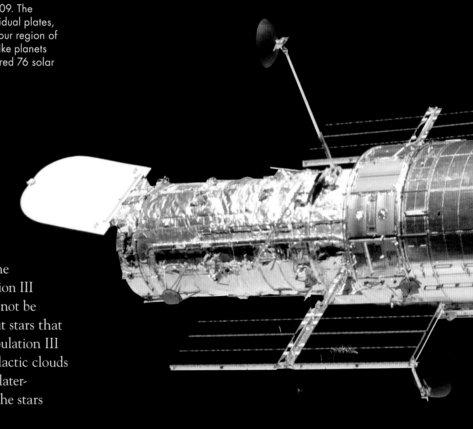

▶ Hubble Space Telescope

Wet planets

Usually only one planet in each system is at the right distance from its central star for it to have liquid water on its surface. It is this subset of wet, warm planets that has most potential to develop life. The spacing of stars in the solar neighborhood indicates that the nearest wet planet is about 25 light-years away. However, given that Earth is the only place we know of with life, we have no way of calculating the chances that other wet planets will also have developed life. We know that life can develop where there is water—since we know that it did so here—but we don't know what other necessary preconditions for life might exist.

Until the mid-1990s our planetary system was the only one we knew of. Then a planet was found orbiting a pulsar. Pulsars are stars at the very end of their life cycles. Because at one time all pulsars were giant stars, it was expected that they would "gobble up" their planets at that stage. Then the star 51 Pegasi was found to have a planetary companion.

Searching for planets

The technique for finding planets is as follows. Imagine that you are looking at the Sun from a nearby planetary system. Jupiter orbits the Sun every 11.9 years, and because Jupiter has a mass that is 1/1047 that of the Sun, it causes the Sun to oscillate about the center of gravity of the Solar System every 11.9 years. In doing this, the velocity of the Sun changes in amplitude by

0.007 mile per second (0.0125 km/s). This produces a Doppler shift. If the star is moving toward the observer, the light's wavelength shifts toward blue; if the star is moving away, it shifts toward red. Modern astronomical spectroscopy is extremely sensitive to small changes in the wavelength of light, and these very small Doppler shifts can now be detected. The observations give the distance between the planet and the star, and the mass of the planet multiplied by a function of the inclination of its orbit. An outside observer with sensitive equipment might detect both Jupiter and Saturn. In addition, an Earth-based observer might have the luck of observing a planetary system edge on. In this case a large planet such as Jupiter may periodically move across its sun's disk. This will reduce the observed brightness by 1 percent, with a transit of measurable duration.

▶ **Distant object**

The Hubble telescope took this image of a distant body orbiting the Sun in the Kuiper Belt. It is Quaoar, a dwarf planet about 800 miles (1,300 km) across, which was discovered in 2002. It is the farthest object in the Solar System ever to be resolved by a telescope, but is too far away to make out details of its icy surface.

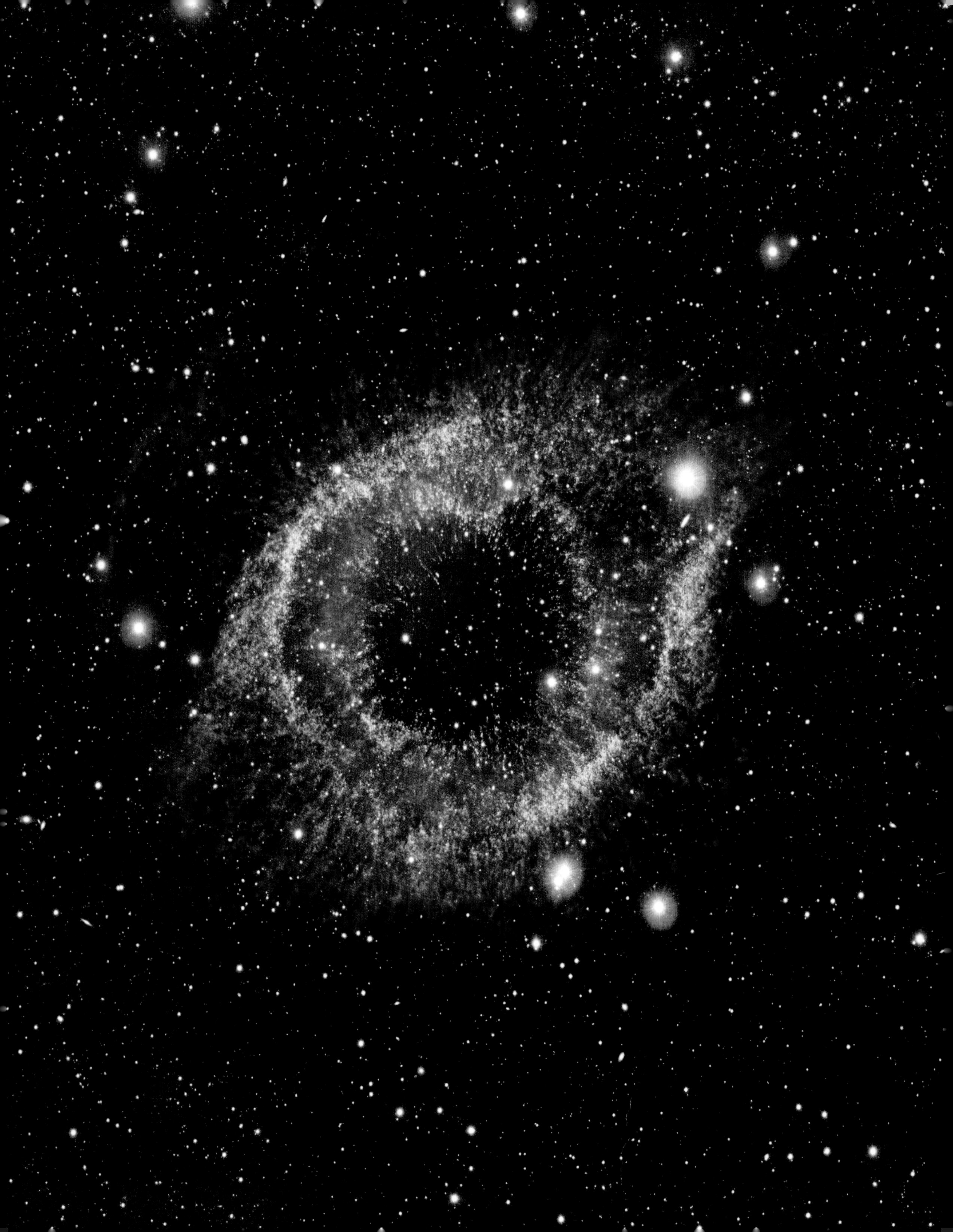

Exoplanet discoveries

More than 200 exoplanets have now been found. The system orbiting the star 47 Ursae Majoris is one of the nearest we have found that is like our Solar System. The two 47 Ursae Majoris planets discovered so far have minimum masses of 2.41 and 0.76 times that of Jupiter, and orbital radii of 2.10 and 3.73 AU (Astronomical Unit—a measurement based on the distance from the Sun to Earth). Most of the other planets that have been found are strange. For example, the planet orbiting the star 51 Pegasi has a mass similar to that of Saturn but is scorchingly close to its star. It orbits every 4.23 days, on a circular orbit of radius 0.05 AU (less than eight times the mean distance between the Sun and Mercury).

Contrary to Copernicus

Findings such as these have left many astronomers confused. For hundreds of years, we have followed what is known as the Copernican Principle. In a nutshell, this says that "there is nothing unusual here, where we live." Earth is not thought to be a specially favored planet, the Sun is expected to be a typical star, and the Milky Way galaxy is thought normal, with plenty like it in the Universe.

Most believers in the Copernican Principle expect that our wet planet, with its collection of continents, oceans, animals, and intelligent humans, is likely to be replicated in many other spots. Carl Sagan, the U.S. space scientist, went so far as to suggest that there are about one million intelligent civilizations in every galaxy of the Milky Way type. More recently, some astronomers have become much more sanguine. Many of the newly found planetary systems are not at all like the one we inhabit. Also, intelligent life seems to be hard to produce and takes a very long time to develop (3.5 billion years in Earth's case). The conditions need to be "just right," and impacts from asteroids and comets affect development. And these are just the variables that we know about—there could be others.

How it will end

The end of the Solar System will be a spectacular affair. In about 5 billion years' time, the fuel in the Sun's core will burn out. Dense shells of helium and hydrogen will produce huge amounts of energy. This activity will make the Sun very unstable, and it will start to successively swell and shrink. Each time it swells, the star's outer layers will be sent out into space. As it sheds material, its exposed surface will grow hotter, and it will glow in different colors. At this point, the Sun will have turned into what is misleadingly known as a planetary nebula.

As the planetary nebula expands, its central star will begin to fade, leaving the exhausted core as a white dwarf. This remnant of the Solar System will be intensely hot, but will appear faint from a distance. By now, the core will have collapsed in on itself to a point where pressure from the electrons will stop it from collapsing any more. The white dwarf left behind by our Sun will be about the size of Earth, but with a density of around 1,000 kg per cm^3— 200,000 times the density of Earth.

Other planetary nebulae

Planetary nebulae were originally named for their spherical, planetlike appearance, but images from modern telescopes such as Hubble have revealed that these beautiful structures come in all kinds of complex shapes. The final shape the nebula will take depends on the interplay of several factors, including the density of the material and the speed at which it was emitted. We have no way of knowing exactly what shape our home star's nebula will take.

▼ **Planets colliding**

Life on Earth could come to an end long before the end of the Solar System—due to a collision with Mercury or Mars. The gravitational tugs that the planets exert on one another could eventually nudge them onto collision courses. Astronomers can only say with confidence that this will not happen in the next 400 million years.

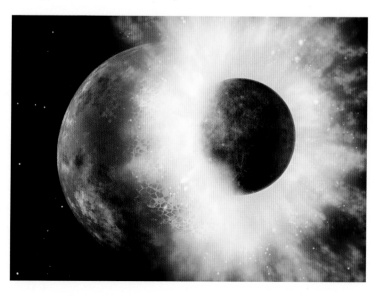

◄ **Helix Nebula**

Captured here by the VISTA telescope, the Helix Nebula is about 700 million light years away. It is one of the closest planetary nebulae to the Solar System. It is the remains of a Sun-sized star, and the core of the nebula is destined eventually to fade into a white dwarf.

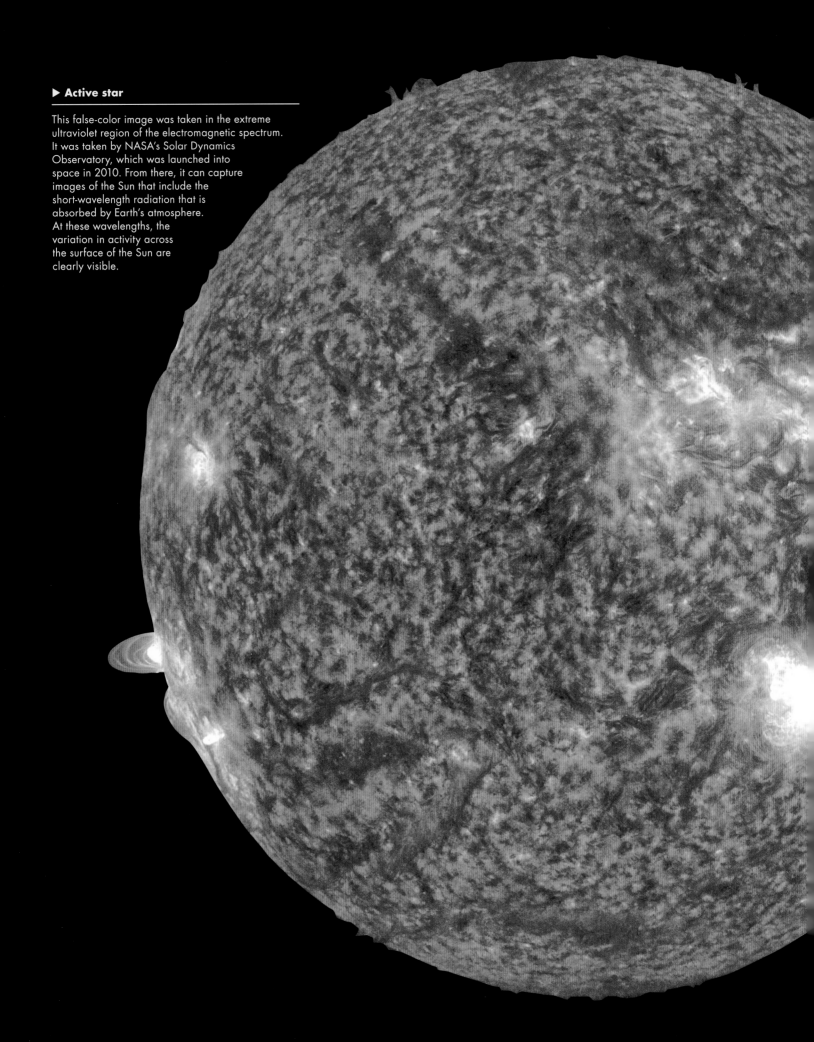

▶ Active star

This false-color image was taken in the extreme ultraviolet region of the electromagnetic spectrum. It was taken by NASA's Solar Dynamics Observatory, which was launched into space in 2010. From there, it can capture images of the Sun that include the short-wavelength radiation that is absorbed by Earth's atmosphere. At these wavelengths, the variation in activity across the surface of the Sun are clearly visible.

THE SUN

The Sun is a fairly typical star, one of at least 200 billion in the Milky Way galaxy alone. It keeps Earth at a temperature above the freezing point of water, even though we are 93 million miles (150 million km) away. For the Sun to maintain this energy output it needs to destroy mass at the rate of 4.3 billion kg/s (kilograms per second). It does this by converting 600 million tons of hydrogen into 596 million tons of helium every second. This might sound like a huge loss of mass, but the Sun has a total mass of 2×10^{30} kg, and when it was first formed, 77.9 percent of its mass was hydrogen. So this mass destruction of the Sun's "fuel" occurs very slowly, and it will last for about 5 billion more years. The energy creation all happens in the central regions of our star, where the temperature is 27 million °F (15 million °C) and the density only about 155 times that of water. At the surface of the Sun, the region that we are able to see, the temperature is a mere 10,000°F (5,700°C) and the pressure about one thousandth that of the atmosphere that we breathe.

▶ The Sun from the ground

When it reaches the top of Earth's atmosphere, sunlight contains about 50 percent infrared light, 40 percent visible light, and 10 percent ultraviolet. The atmosphere blocks most of the high-energy ultraviolet radiation, so that only around 3 percent of the light that reaches the ground is ultraviolet. This is important for the survival of life at ground level. It is unlikely that life could have evolved on land with high levels of ultraviolet radiation in the sunlight, as it damages living cells.

◀ Corona

During a total solar eclipse, the Moon covers the photosphere, revealing the surrounding corona as a bright halo. The ionized particles in the corona are shaped by the Sun's magnetic field, giving the corona a somewhat spiky structure. The pink patches just visible around the silhouette of the Moon are part of the Sun's chromosphere.

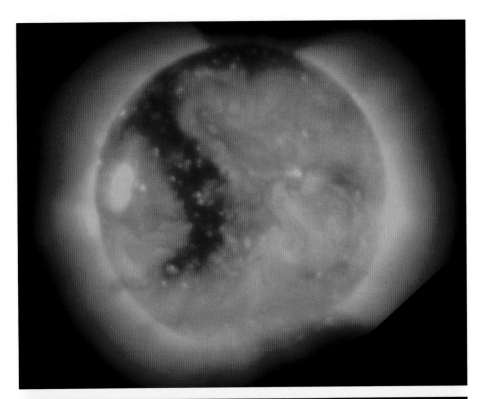

Constant presence

Analysis of the evolution of our planet, its geology, and its life forms indicates that the temperature has remained very nearly constant throughout its existence. This means that the rate at which energy has been given out by the Sun has changed very little throughout the Sun's life. Both the Sun and Earth are about 4.6 billion years old, and were formed out of the same cloud of gas and dust. As it has emitted energy constantly throughout this period, the Sun has destroyed a mass of 6×10^{26} kg. This is 8,000 times the mass of our Moon, but only about 0.03 percent of the Sun's original mass. Since the solar mass produces the gravitational pull that keeps Earth in orbit, this loss of mass means that our planet is now about 30,000 miles (50,000 km) farther away from the Sun than it was originally.

Changing composition

As its hydrogen is converted into helium, the composition of the Sun slowly changes. When it first formed, it was 77.9 percent hydrogen, 20.4 percent helium, and 1.7 percent other elements. Scientists loosely refer to everything that is not hydrogen and helium as "metal." The Sun now has a composition of 73.4 percent hydrogen, 24.9 percent helium, and 1.7 percent metal. Not all the hydrogen is available as fuel, and it has been estimated that the Sun has about five billion years to go before usable hydrogen runs out.

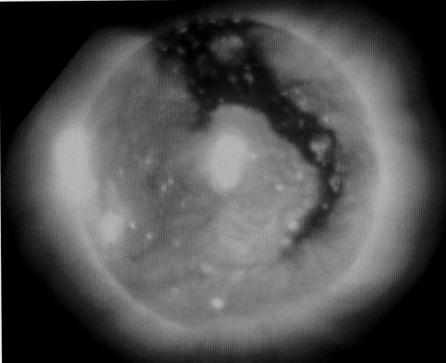

▲ **Coronal hole**

These X-ray photographs were taken about two days apart by the Skylab space station. They reveal a boot-shaped hole in the corona that changes position as the Sun rotates. Coronal holes create high-velocity streams in the solar wind. These streams affect Earth's magnetic field and disturb Earth's ionosphere, possibly changing weather patterns.

Internal structure

The visible surface of the Sun, where energy escapes in the form of radiation, is called the photosphere. Beneath this lie three distinct layers. At the center is the core, where temperatures and pressures are high. Here, nuclear fusion turns the protons of hydrogen into helium nuclei at a rate of about 600 million tons per second. The energy produced by this process is emitted in the form of radiation and neutrinos. The energy travels out into a cooler region called the radiative zone. It takes about 1 million years to move through the radiative zone, as the radiation is repeatedly absorbed and re-emitted by the plasma (ionized gas). Farther out, the plasma starts to flow in currents, as hot plasma wells up against cooler plasma that is falling. This is the convective zone, which carries the energy to the surface.

Uneven surface

The photosphere is a layer of plasma about 60 miles (100 km) thick. It has a granulated appearance with bumps about 600 miles (1,000 km) wide. The bumps are the upper surfaces of convection cells bringing hot plasma up from the interior.

Magnetic field disturbances

The Sun is a rotating body made up mostly of electrically charged particles. This combination produces strong magnetic fields. The convective zone rotates faster at the equator than at the poles, which causes the magnetic field lines to become twisted. Where the magnetic field lines are concentrated, the flow of heat from the interior is slowed down, which produces cooler areas in the photosphere. The cool areas appear as dark sunspots.

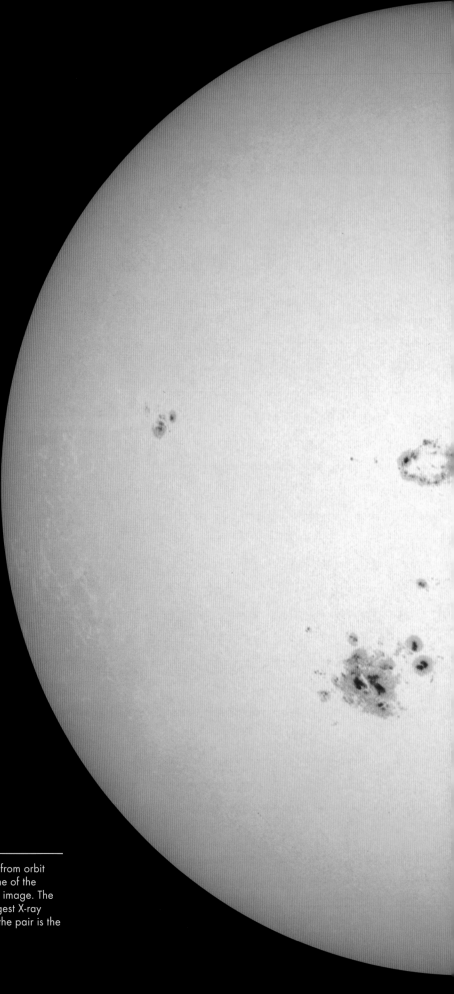

▶ **Huge sunspot**

The Solar and Heliospheric Observatory (SOHO) observes the Sun from orbit around Earth. It took this image on October 28, 2003, capturing one of the largest sunspot groups ever observed, seen here in the center of the image. The spot occupied an area equal to 15 Earths and later fired off the largest X-ray flare ever recorded. Sunspots usually appear in pairs; each spot in the pair is the opposite magnetic pole to the other.

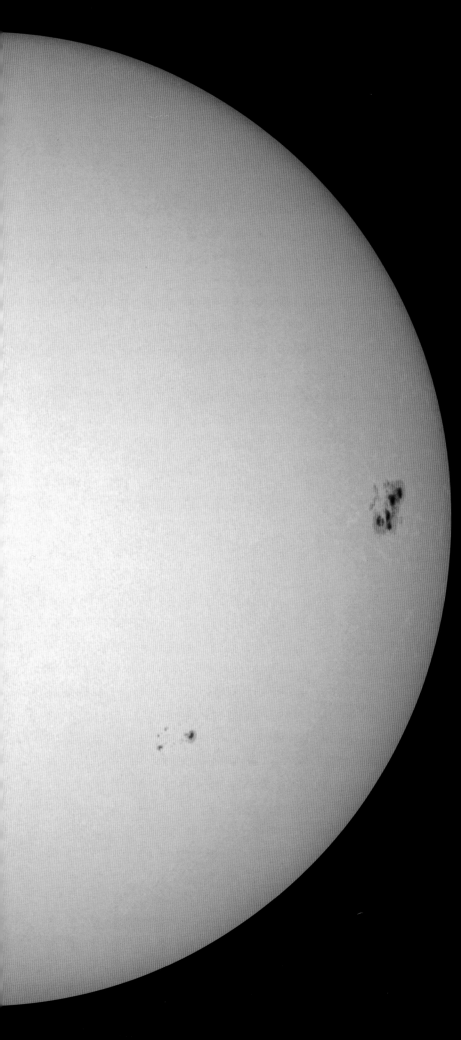

▼ **Sunspot group**

A large group of sunspots was observed on September 23, 2000. It covered an area about 12 times larger than the surface of Earth. Each spot has a dark central area called the umbra, and a lighter outer area called the penumbra. The temperature at the center of the spots is typically about 4,900–7,600°F (2,700–4,200°C), compared to 10,000°F (5,500°C) in the surrounding area. Although the spots appear dark against the surrounding bright area, if you were able to isolate a sunspot and place it in the night sky, it would still be brighter than the Moon.

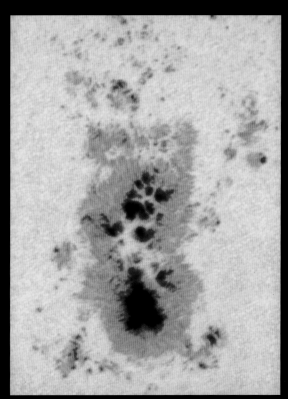

Cycle of activity

Other phenomena such as solar flares and plasma loops are also caused by disturbances in the Sun's magnetic fields. Where plasma is heated along magnetic field lines, it can explode up into the Sun's atmosphere as solar flares. Twisted magnetic field lines can also cause the eruption of loops of plasma. Prominences of dense clouds of plasma, suspended above the surface by loops in the magnetic field, can endure for several weeks. The level of activity in the photosphere moves from a minimum to a maximum over an 11-year cycle.

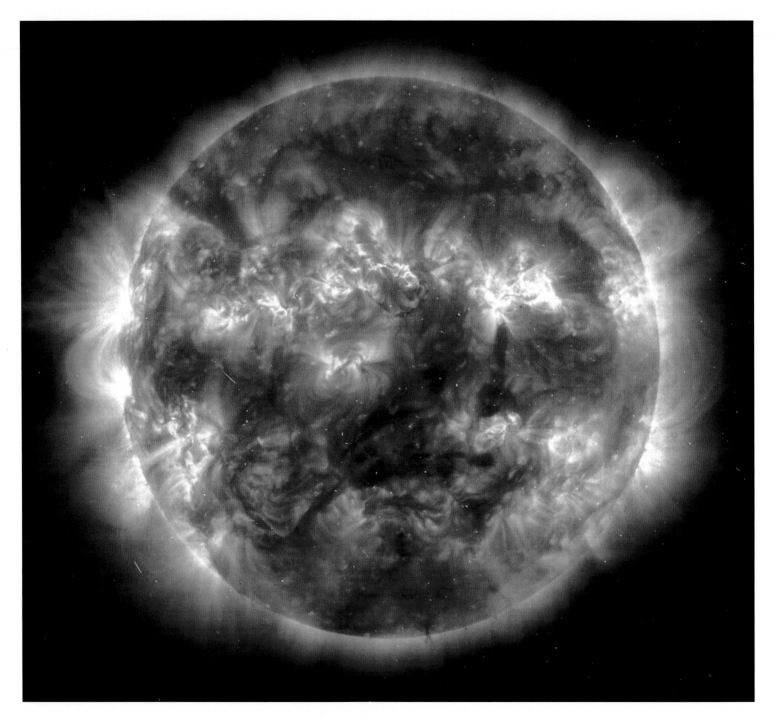

▲ **Near solar maximum**

This SOHO false-color image was taken in 2000, around a solar maximum, when the Sun was peppered by active regions. The white specks are masses of charged protons, which were released a few hours before this image was taken in a coronal mass ejection.

The Sun's atmosphere

Above the photosphere, which forms the lowest layer of its atmosphere, are three more layers. The red-orange chromosphere extends about 1,200 miles (2,000 km) above the photosphere. The temperature rises through this layer, and at the top it is about 36,000°F (20,000°C). Above the chromosphere is a thin layer called the transition layer. Here, the temperature soars to 1.8 million °F (1 million °C), for reasons that are poorly understood. Above this layer is the Sun's corona, which is a layer of thin plasma. The corona is even hotter, at 3.6 million °F (2 million °C). It extends millions of miles into space, eventually blending with the solar wind, a stream of charged particles that flows from the Sun deep into the Solar System.

Metal-rich star

When the age of the Sun (4.6 billion years) is compared with the age of the universe (14 billion years), it becomes clear that the Sun is very much a second-generation star, born containing many metallic remnants from massive stars that have "died" and become supernovae. This is important when it comes to the existence of planets. Even when compared with other stars of similar mass and age in the same region of our galaxy, the Sun appears metal-rich and therefore much more likely to have planets than its neighboring stars.

Even though only 1.7 percent of the Sun's mass is metal, this is 5,000 times the mass of Earth. The most abundant metal in the present-day solar photosphere is oxygen (1 percent), followed by carbon (0.3 percent), neon (0.2 percent), and iron (0.2 percent). Almost all of these metals were already present in the interstellar medium from which the Sun formed.

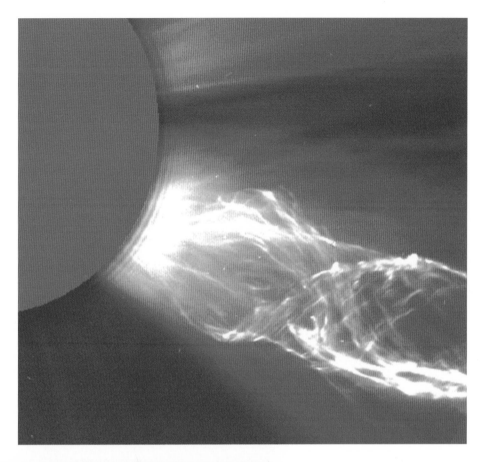

▲ Helical ejection

This helical coronal mass ejection was captured by the LASCO C2 coronagraph on June 2, 1998. To take the image, direct sunlight was blocked by an occulter, revealing the faint corona. The Sun is approximately half the diameter of the red circle.

◄ Helium prominence

An erupting prominence was seen on July 1, 2002. This image of the prominence was taken using a filter that highlights ionized helium at temperatures of about 110,000°F (60,000°C).

The solar cycle and Earth

At the peak in a cycle of solar activity, sunspots appear about 40 degrees north and south of the equator. As activity declines, they move closer to the equator. Very few spots were recorded between the years 1645 and 1715, a cold period known as the "Little Ice Age," and it has been suggested that variations in the solar cycle have an effect on Earth's climate. The total output of solar energy varies little through the cycle, but its X-ray and ultraviolet emissions do fluctuate. This energetic short-wavelength radiation interacts with Earth's ionosphere in the upper reaches of the atmosphere, and may influence the formation of clouds.

It has also been proposed that, in addition to the 11-year cycle, there are longer-term cycles in solar activity, caused by instabilities in the Sun's core. These fluctuations may repeat with a period of around 100,000 years, which would explain the regular periodicity of ice ages on Earth.

To better understand the effects of these changes in solar activity, NASA observes the Sun from observatories in space, where it is possible to take images of our home star in a wide variety of wavelengths without interference from Earth's atmosphere, which blocks much of the X-ray and ultraviolet radiation. Further research is needed to work out exactly how solar activity affects us.

▶ **Aurora borealis**

Charged electrons and protons stream out from the Sun in the solar wind. When they reach Earth, some of these particles become trapped over the magnetic poles. Here, they collide with atoms in the ionosphere, producing spectacular light shows known as auroras. The aurora seen over the North Pole is known as the aurora borealis, or Northern Lights.

▼ **Aurora on Jupiter**

The solar wind produces auroras over the poles of other planets. In this image taken in 2004, the Hubble telescope has captured a curtain of glowing gas wrapped around Jupiter's north pole like a lasso. The image was taken in ultraviolet light, and also shows the glowing "footprints" of three of Jupiter's largest moons: Io, Ganymede, and Europa.

◄ The smallest planet

The smallest planet in the solar system, Mercury is also the closest planet to the Sun. With only a very thin atmosphere, and a slow rate of spin, it has long days and nights, making the surface alternately hot enough to melt lead or cold enough to liquefy air. This image reveals the surface of the planet to be very similar to our Moon's, with many impact craters and features shaped by ancient lava flows.

MERCURY

Mercury is tiny, having a mass only 5.5 percent that of Earth. Knowing its average density, we realize that, like Earth, it is divided into an inner metallic core and a rocky outer mantle. But out of its total radius of 1,516 miles (2,440 km) the rock mantle is only 342 miles (550 km) thick. Mercury's low mass means that it has a small gravitational field, and that much more material escapes from its surface when it is cratered than occurred on Earth.

A considerable amount of the outer mantle may have been lost during the early days of the Solar System, either through extreme heat in the solar nebula or through asteroidal bombardment.

▼ **MESSENGER to Mercury**

Only two spacecraft have visited Mercury. Mariner 10 flew past in 1974–75 and imaged about 45 percent of the surface. By 2012, NASA's robotic Mercury Surface, Space Environment, Geochemistry, and Ranging (MESSENGER) spacecraft was in orbit around the planet, having made three flybys in 2008–09. The craft continues to photograph the surface. It has neutron, gamma-ray, and X-ray detectors to study the crust, a laser altimeter to produce an accurate contour map, and an ultraviolet spectrometer to measure the composition of the atmosphere.

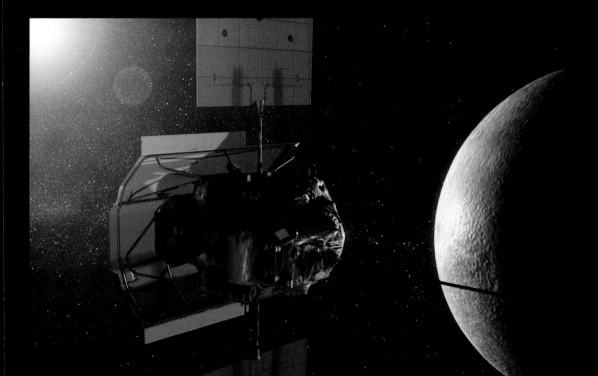

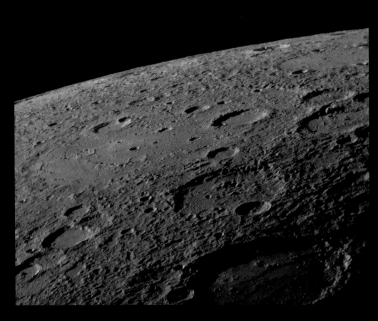

▲ Surface details

As the MESSENGER spacecraft drew closer to Mercury for its historic first flyby in 2008, the narrow angle camera took this image of the sunlit side of the planet. It shows a variety of surface textures, including smooth plains, many impact craters (some with central peaks), and rough material that may have been ejected from the large crater to the lower right.

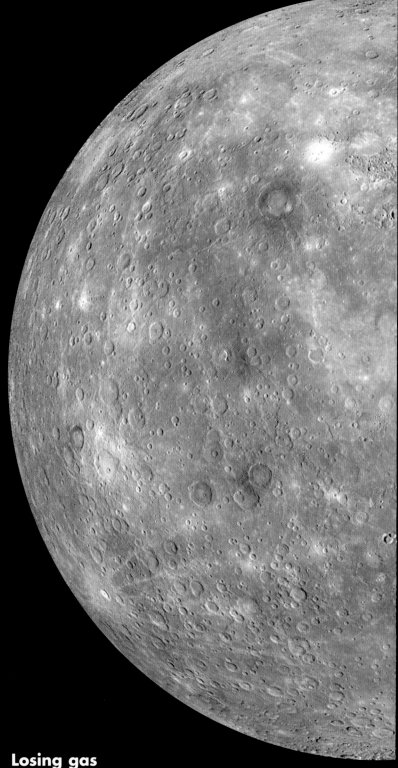

Slowly spinning planet

Because Mercury is the innermost planet, it never moves more than 28 degrees away from the Sun in the sky and so its features are very hard to see from Earth. In the late 19th century, astronomers estimated the planet's spin period as 88 days, the same as its orbital period. They thought that Mercury was in a 1:1 synchronous spin–orbit resonance, just like the Moon. If this was so, one face would point permanently to the Sun and be scorching hot and the other side would be frigid, always dark, and pointing out to space. By looking at radar pulses reflected from the surface, scientists in the 1960s realized that their predecessors were wrong. Mercury is spinning every 56.6462 days. It spins three times in every two orbits of the Sun, so one day on Mercury, from sunrise to sunrise, is 176 days, nearly twice the length of the Mercury year. The temperature at the equator at noon is a scorching 806°F (430°C). At midnight it, drops to –292°F (–180°C).

Losing gas

Mercury has a very tenuous atmosphere of oxygen, sodium, hydrogen, and helium. It is continuously being lost and then replenished because the gravitational field is too small to hold on to it. Interestingly, the temperature difference means that the atmosphere is much denser on the night side than on the day side. The molecules are forced out of the surface by impacting ionized particles from the magnetic fields around Mercury, and some molecules derive from the wind of gas escaping from the Sun.

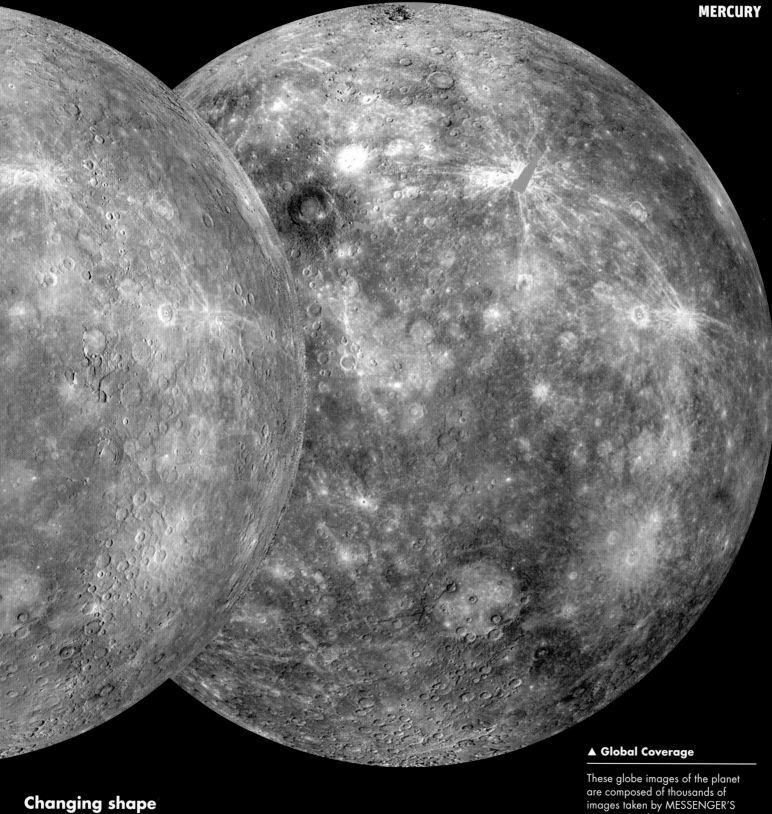

Changing shape

Mercury was formed by the accretion of rocky planetesimals.
At first it would have had a spin period of less than 24 hours.
Friction slowed it down within 500 million years of its
formation. Planets that spin quickly are obloid in shape,
like a squashed ball. Slowly spinning planets, however, are
spherical. The slowing of Mercury's spin changed its shape
over time and disrupted the surface.

▲ **Global Coverage**

These globe images of the planet
are composed of thousands of
images taken by MESSENGER'S
Mercury Dual Imaging System
(MDIS) in 2011. The globe on the
right was created from the MDIS's
wide-angle camera using its red,
blue, and green filters. The result is
how the planet would look to the
human eye. The camera system
provides a complete map of the
surface of Mercury at a resolution
of 820 ft (250 m) per pixel.

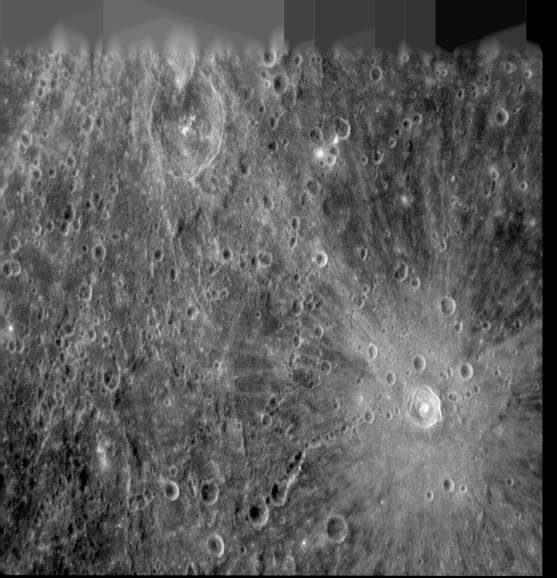

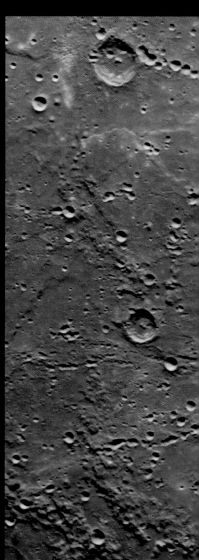

◄ Crater discovery

During its flyby of Mercury of January 14, 2008, MESSENGER took this image of a previously unseen crater with distinctive bright rays of ejected material extending outward from the center. A chain of craters nearby is also visible. Studying impact craters gives us an insight into the history and composition of Mercury, as well as of events that occurred throughout the solar system.

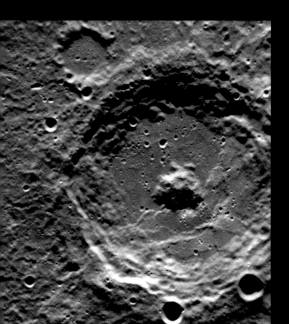

◄ Donne crater

This image from the narrow angle camera provides a fine view of the crater Donne. (John Donne was an English poet known for his sonnets and elegies.) Donne crater displays a mound-like central peak that has been somewhat rounded by impact erosion, and well-developed wall terraces. The crater has been modified by lobate scarps (lobe-shaped cliffs, or ridges) that cut the walls and floor.

Cooling down

Mercury's small size has allowed it to cool quickly. Most of the core is probably solid, and the crust too thick for continental drift. Scientists have calculated that the cooling reduced the radius of the planet by about 1.2 miles (2 km). The cooling process would also have disrupted the shape of the surface.

Moon-like surface

Mercury is covered in impact craters. Unlike our Moon, however, the ejecta blankets are closer to the parent craters, and are thicker. This is because the planet's gravity is about twice that of the Moon's. Large meteorite impacts have produced multi-ring basins, the most impressive of which is the Caloris basin. On the opposite side of the planet to the basin is a region of strange terrain produced by earthquakes resulting from the impact. The craters are interspersed by at least two generations of flat plains of solidified lava, like the luna maria. Lava oozed from vents in the crust and pooled into depressions. A unique feature is a series of cliffs that runs from north to south. These were probably formed as the planet cooled and was forced to spin more slowly.

▼ Double ring crater

This scene was imaged by MESSENGER's narrow angle camera during the spacecraft's flyby of Mercury on January 14, 2008. The outer diameter of the large double ring crater at the center is about 160 miles (260 km). The crater is filled with smooth plains material that may be solidified lava. Multiple chains of smaller secondary craters are also seen extending radially outward from the double ring crater. Double or multiple rings form in craters with very large diameters, often referred to as impact basins.

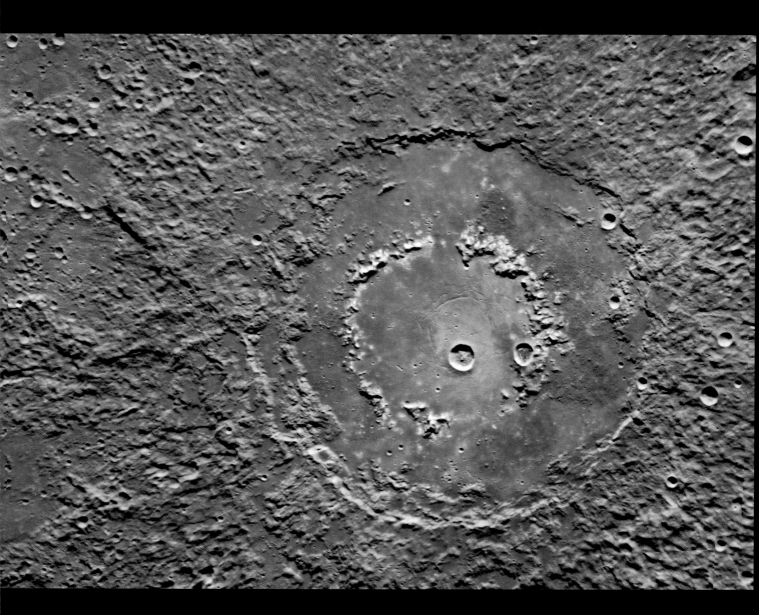

▼ Mercury's Caloris Basin

The Caloris Basin appears as a large orange circle in this false-color mosaic taken by MESSENGER in its January 2008 flyby. The orange splotches around the basin's perimeter are thought to be ancient volcanic vents, new evidence that Mercury's smooth plains are indeed solidified lava flows.

Caloris Basin

One huge impact basin, named Caloris, is located in the northern hemisphere and is 832 miles (1,340 km) in diameter (27 percent of the diameter of the planet). It is larger than the state of Texas and is the second-largest impact crater in the inner solar system, the largest being the South Pole–Aitken crater on the Moon.

The basin is named "Caloris"—Latin for "heat"—because it faces the Sun at perihelion and becomes one of the hottest places on the planet. The asteroid responsible for creating the basin was huge—probably 90 miles (150 km) in diameter. Ejecta were thrown more than 620 miles (1,000 km) beyond the outer rim, producing many radial ridges. This dramatic event probably happened towards the end of the early Solar System's period of massive bombardment. The basin's floor later filled with lava, which cooled and fractured in a polygonal fashion. The basin is now about 1.2 miles (2 km) deep. Shock waves from the Caloris impact traveled around the planetary crust, producing earthquakes and massive upheavals across 96,500 square miles (250,000 square kilometers) on the opposite side. These were then reflected back to the basin and fractured the surrounding rocks.

Future exploration

In addition to mapping the surface comprehensively, MESSENGER has expanded our knowledge of Mercury with several important new discoveries. There is evidence, for example, that Mercury, like Earth, has a global magnetic field generated by a dynamo process inside its core, which is very large for the planet's size—about 85 percent of its diameter—and remains at least partially liquid, not fully solid as was previously thought.

But perhaps most surprisingly for the planet nearest to the Sun, Mercury is now known to have ice inside shaded craters near the north pole, with the possibility that there may even be liquid water beneath some of the permanently shadowed regions. In time, these may provide clues as to how life began on Earth.

▶ Mercury's Caloris Basin

This is a computer photomosaic close-up of the Caloris Basin, showing many generations of impact craters. It was taken by the Mariner 10 spacecraft during its flyby of Mercury. The spacecraft took images of Venus in February 1974 on the way to three encounters with Mercury in March and September 1974 and March 1975. The spacecraft took more than 7,000 images of Mercury, Venus, Earth, and the Moon during its mission.

VENUS

Venus is often referred to as Earth's twin. Its mass and diameter are 81.5 percent and 95 percent those of Earth, and its rocky-metallic composition is very similar, the average density being 5.2 times that of water. The significant difference is that Venus is 28 percent closer to the Sun. This has made it hotter and has driven away the water, with two major consequences. The carbon dioxide that was released by the cracking of the rocks has not been converted into chalk, but remains in the atmosphere. In addition, the lack of water in the crust has dramatically increased the viscosity of the underlying mantle. The crust has not broken into plates, like Earth's crust.

From Earth, Venus appears as the brightest planet, but it can be seen only as what is called the Evening, or Morning, Star. It never travels more than 47 degrees from the Sun so is always low in the sky. It is a beautiful sight, but all we can see on it are featureless clouds. These form an all-encompassing layer that blankets the whole planet. They start 28 miles (45 km) above the surface and are about 16 miles (25 km) thick.

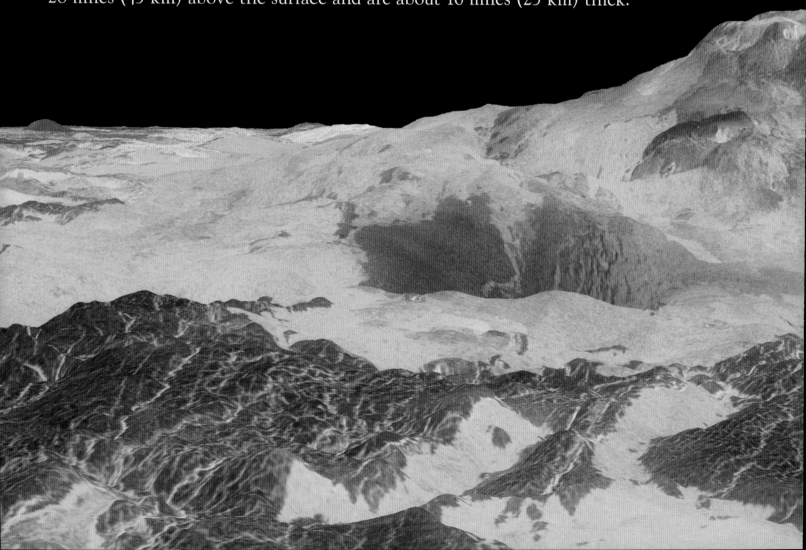

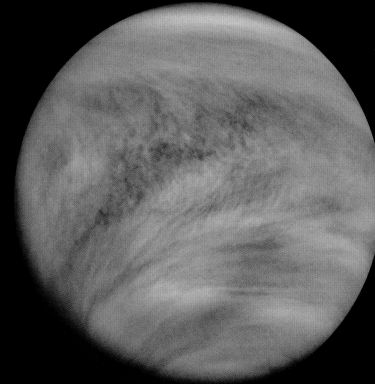

◄ Cloudy planet

Venus's thick atmosphere was photographed in ultraviolet light by the Pioneer Venus Orbiter on February 5, 1979. Pioneer Venus used an orbiter and several small probes to study the planet both from above and from within the clouds. The orbiter collected information about Venus for more than a decade until it finally ran out of fuel and was destroyed on entering the atmosphere in August 1992. Information collected by Pioneer Venus was used to produce the first global topographical map of the surface of the planet, which is permanently hidden from view by its atmosphere.

▼ Venusian volcano

This computer-generated view of the surface of Venus was created using radar data from NASA's Magellan spacecraft and shows the planet's largest volcano, Maat Mons, which rises 3 miles (5 km) above the surrounding ground. Lava flows from the volcano hundreds of miles across the fractured plains in the foreground. The vertical scale in the image has been exaggerated 10 times, and the mountain has only gently inclined slopes. The colors are based on images recorded by the Soviet Venera 13 and 14 spacecraft, which landed on Venus in 1981.

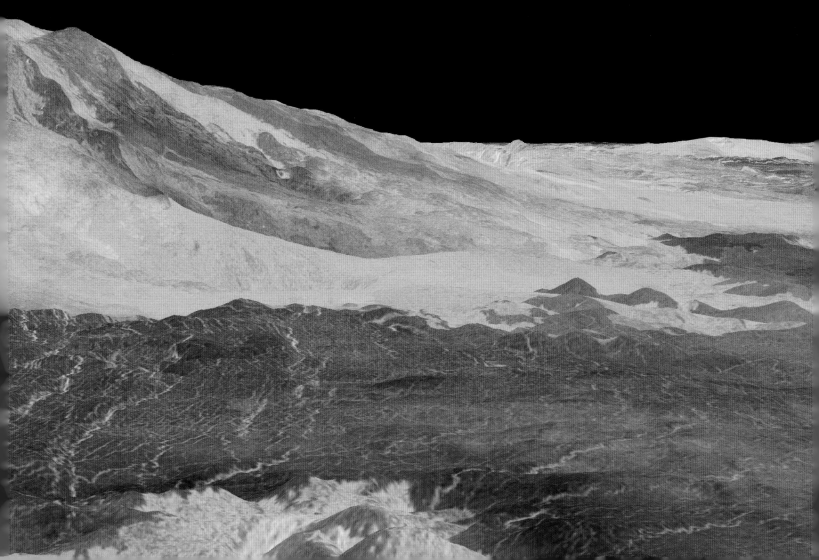

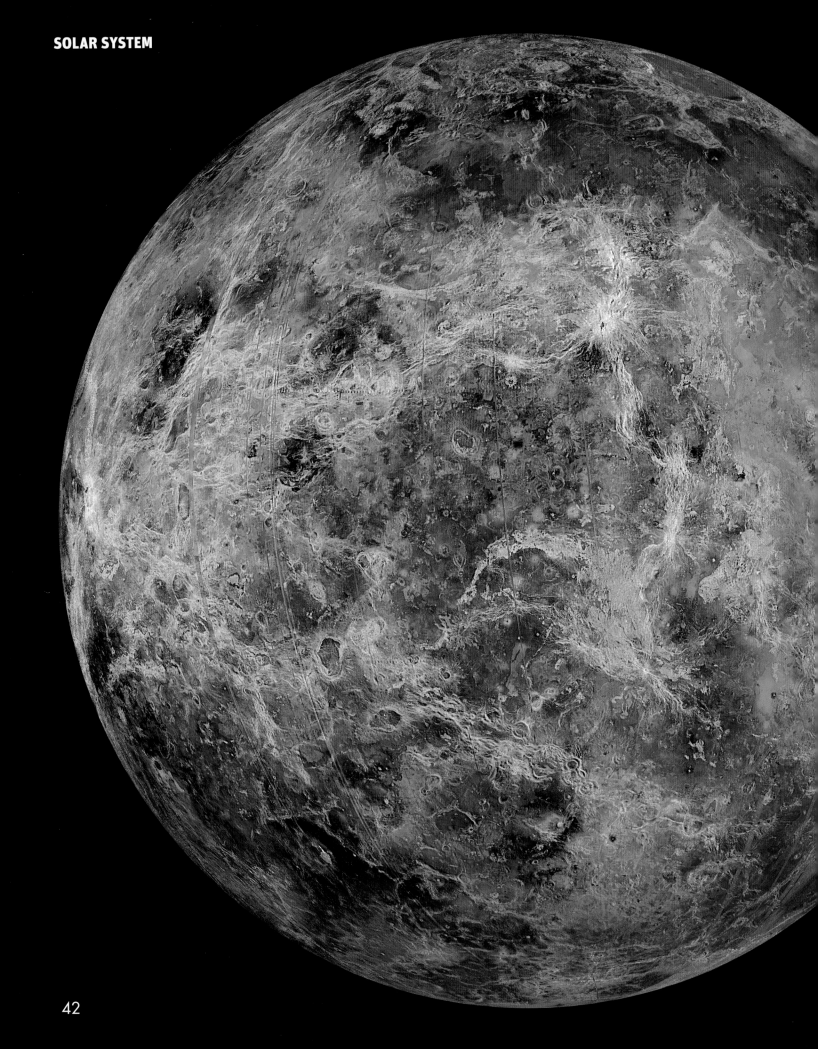

Surface conditions

Eighty percent of the incoming sunlight is reflected by the cloud tops of Venus, which consist mostly of droplets of sulfuric acid. The 20 percent of light that does make it through the clouds illuminates the overcast ground with a dull orange light. In the 1970s, the Soviet Venera probes made controlled landings on the surface, sending the first images from the surface in 1975. They found that the atmosphere is 96.5 percent carbon dioxide and about 3.5 percent nitrogen. The pressure is about 92 times that at the surface of Earth. The average temperature at the surface is 867°F (464°C), and there is very little difference in conditions between the equator and the poles. The temperature also remains relatively constant between day and night. There is a slight breeze toward the poles, but essentially no weather. Prevailing winds have, however, produced streaks and dunes of dry, dusty sand around prominent craters.

Trapped heat

The carbon dioxide in the atmosphere generates a massive greenhouse effect. The visible light from the Sun can penetrate it, and provides heat, but the infrared radiation given off by the hot ground is trapped. The equatorial plane is tilted by only 2.6 degrees to the plane of the orbit, so there are no seasons. Ascending through the atmosphere, the temperature decreases by about 20°F (11°C) per mile. This explains why it is cold enough to form clouds at a height of 28 miles (45 km), where the temperature drops to below 300°F (150°C). There are three distinct cloud layers between 23 and 43 miles (45 and 70 kilometers) above the surface.

◀ The surface of Venus

This false-color hemispheric view of Venus is centered at 270 degrees east longitude. The image has been put together using a number of images taken by the Magellan spacecraft. Magellan has imaged 98 percent of the surface of Venus to a resolution of about 300 feet (100 meters). Gaps in the data were filled in with information from the earlier Venera and Pioneer Venus missions.

▶ Magellan is deployed

Seen here being deployed from the Space Shuttle Atlantis on May 4, 1989, the Magellan probe reached orbit around Venus in August 1990, and remained there for the next four years, mapping the surface of the planet. At the end of its mission, Magellan's orbit was lowered, and the craft sent back valuable information as it moved toward the surface. It lost contact on October 12, 1994, when it is believed to have burned up in the atmosphere near the surface.

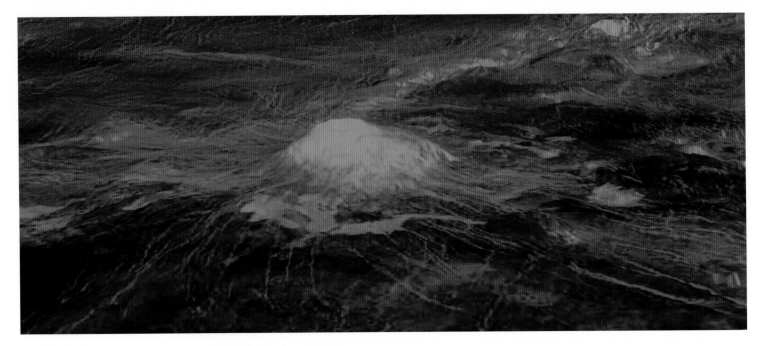

▲ **Iduun Mons**

This image of Iduun Mons volcano, put together from data gathered by NASA's Magellan probe, has a vertical exaggeration of 30 times. Bright areas are rough or have steep slopes. Dark areas are smooth.

▶ **Alpha Regio**

Alpha Regio is an upland area about 800 miles (1,300 km) across. In 1963, it was the first feature on the surface of Venus to be identified from an Earth-based radar. Its center is 25 degrees south latitude, 4 degrees east longitude. It contains structural features including ridges, troughs, and flat-floored valleys. Directly to the south of the complex, ridged terrain of Alpha Regio is a large, ovoid-shaped feature called Eve. The bright spot at the center of Eve is the point chosen to act as the prime meridian of Venus when making map references. The vertical scale of the map has been exaggerated about 23 times.

Landing on Venus

The Soviet Union's Venera missions were the most successful probes to investigate Venus. Sixteen craft were launched over a period of 20 years, making the first soft landing on the planet. The Venera 4 craft was the first to send back information, making measurements of the atmosphere in October 1967. On December 15, 1970, Venera 7 became the first craft to land on the surface, sending back information for 23 minutes before its batteries ran flat. Subsequent Venera missions were able to study the surface in detail and sent back photographs. Veneras 13 and 14 were fitted with drilling arms and were able to take samples, revealing the ground to be basalts similar to those found on Earth. The final two missions, Veneras 15 and 16, were orbiters, and worked together for nine months from October 1983 to scan the surface using radar from its north pole down to about 30 degrees north.

First maps

The information sent back by NASA's, Pioneer Venus spacecraft, launched in 1978, was used to produce the first maps of the surface. Two major continental massifs were seen rising up from the surrounding lowland. These were called Ishtar Terra and Aphrodite Terra. The Venusian topography is less varied than Earth's. About 60 percent of the land is within 1,600 feet (500 m) of the mean radius. Eighty-five percent of the surface is basaltic plain.

Completing the survey

In the early 1990s, the U.S. Magellan probe made the first detailed survey of the surface with a hugely improved resolution. The craft completed one orbit every 3 hours 15 minutes, moving between the north and south poles. Venus turns once in 243 Earth days, and as the planet slowly turned under the probe, it was able to map the surface strip by strip. After one complete spin of the planet, the mapping was complete. Two more mapping cycles followed, producing detailed maps of 98 percent of the planet's surface. The look angle of the radar differed from cycle to cycle, and the slightly different views of the same location were combined to produce three-dimensional images. Magellan revealed a volcanic landscape dominated by ancient lava and marked by impact craters.

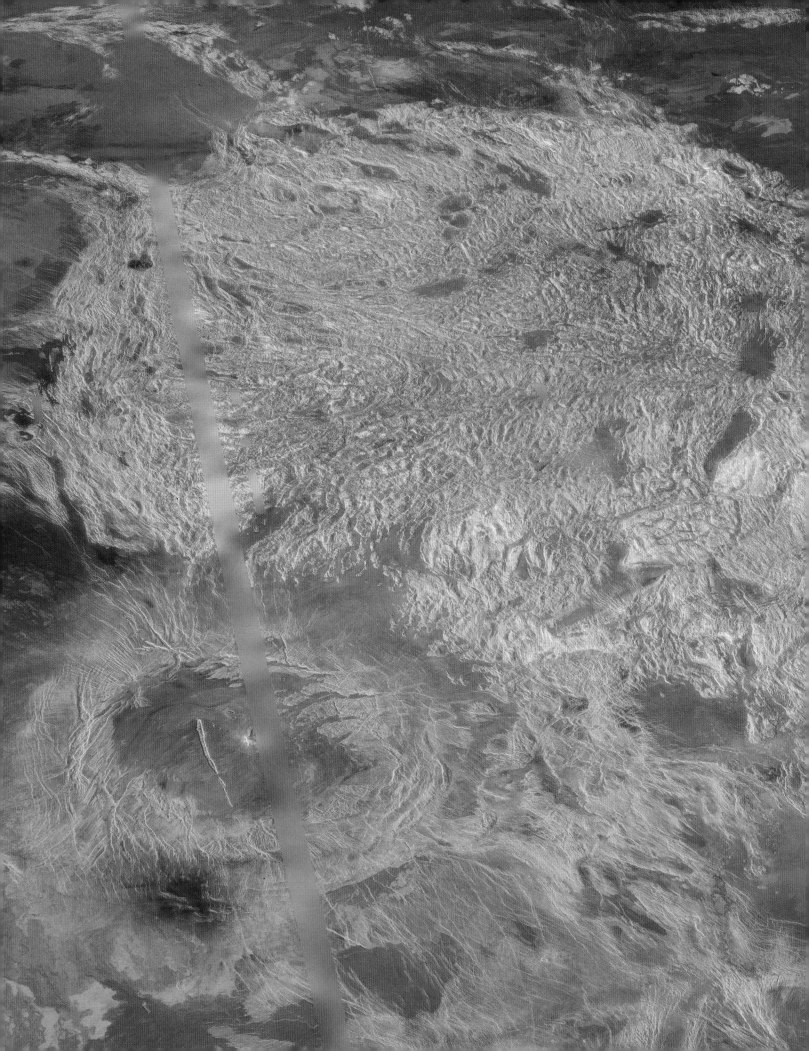

Volcanoes

There are hundreds of volcanoes on Venus, many with gently sloping sides like the shield volcanoes of Hawaii. Maat Mons, the biggest, rises 3 miles (5 km) above the plains and is 5 miles (8 km) above the mean surface level. Unlike most of Earth's volcanoes, which are distributed along the edges of subducting plates, the volcanoes on Venus are randomly dotted over the planet's surface. The lava floodplains are mostly between 500 and 700 million years old, so in terms of normal planetary timescales, the planet has been relatively recently resurfaced. Some of the volcanoes are probably still active, but the activity will be much less than it was in the past. The Venera spacecraft that touched down on the surface recorded data for about 60 minutes before being overcome by heat and acidity. The atmosphere near the surface was found to be free of haze. There was little soil but many layers of dark basaltic rock.

Craters on Venus

Venus is named after the Roman goddess of love, and its craters have also been named after women. Hundreds of impact craters have been discovered, ranging in diameter from 0.6 to 168 miles (1 to 270 km), but the number is smaller than exist on equivalent areas of the Moon. Small craters are not produced on Venus because the atmosphere erodes the incoming impactors, and the recent resurfacing has erased all the craters that were formed during the planet's early history. The depth of the craters and the height of their walls are less than in craters of similar size on the Moon. The crust's warmth means that it has less strength, and the craters show viscous relaxation.

Slow spin

Venus spins once every 243 days, 36 minutes. The spin is retrograde, which means that it is in the opposite direction to that of Earth. How and when Venus was slowed down are mysteries. It is also not understood why the upper atmosphere whizzes around the planet once every four Earth days. The European Space Agency's Venus Express, launched in 2005, is observing the atmosphere to better understand its dynamics.

One consequence of Venus's slow spin is that the liquid core of the planet does not produce a detectable magnetic field. Seismometers have not yet been placed on the surface of Venus, so we are not sure about the extent of the core.

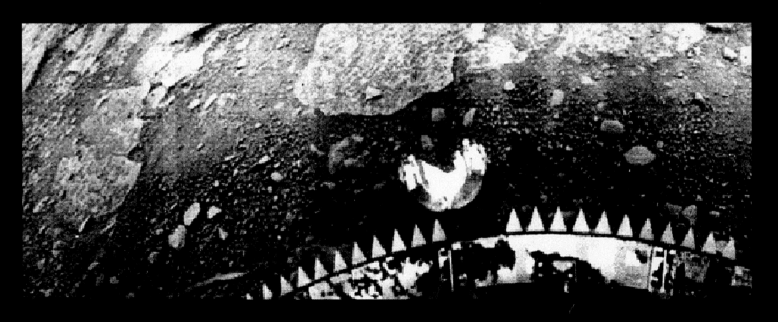

EARTH

The presence of liquid water makes Earth different from the other planets. Our planet is at just the right distance from the Sun to be wet. Venus is too close and too hot, while Mars is too far away and its water has frozen or gone underground. Water covers 75 percent of Earth's surface, to an average depth of about 1.8 miles (3 km), and lubricates the plate tectonic process by reducing the viscosity of lava. About 97 percent of Earth's water is in the oceans at any one time. Of the remaining 3 percent, two-thirds is locked up as glaciers and ice sheets and one-third is in the form of rivers, lakes, groundwater, and clouds. Even though water dominates our existence, it actually makes up only one part in 4,800 of Earth's mass. The atmosphere, with its mass of 5.4×10^{18} kg, accounts for a mere one part in 1,100,000.

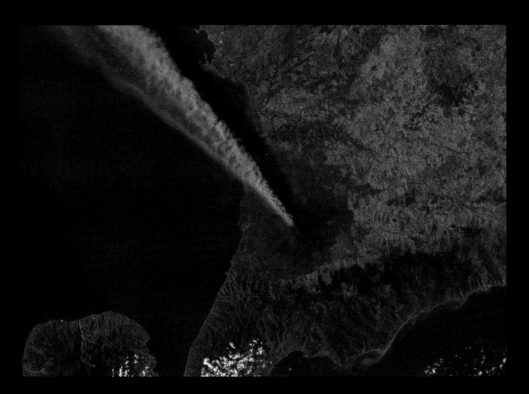

▶ Blue Marble

This photograph of Earth was taken on December 7, 1972, from Apollo 17. The spacecraft was about 28,000 miles (45,000 km) away from the planet on its way to the Moon. The Sun was directly behind it, so the whole facing hemisphere of the planet is lit up. This was the first time the Apollo trajectory had made it possible to photograph the south polar ice cap. In the top right of the image, a cyclone is crossing the Arabian Sea. The cyclone had ravaged Tamil Nadu in India two days earlier.

▲ Active volcano

Records of frequent eruptions from Mount Etna, on the island of Sicily in the Mediterranean, date back to 1500 BCE. This photograph of an eruption was taken on September 3, 2010, from the International Space Station. There are two plumes of smoke—a dense plume from one of the three summit craters, and a lighter plume from an eruption on the mountain's flank.

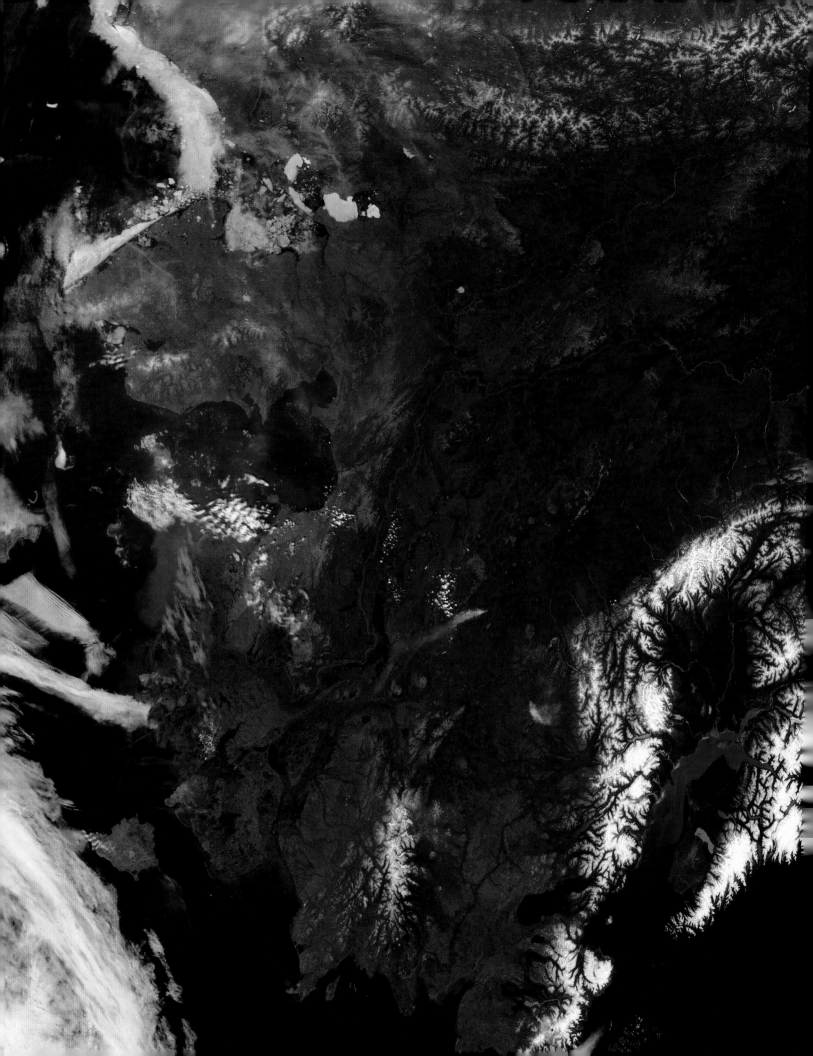

Tectonic plates

Earth is the only terrestrial planet whose crust has broken into tectonic plates that move around. This makes Earth much more active than its rocky neighbors. The plate movements result in the production of huge mountain ranges such as the Himalayas and the Rockies. The top of Mount Everest in the Himalayas is 5.5 miles (8.8 km) above sea level and is rising by a few fractions of an inch per year because the Indian plate is being pushed underneath the Eurasian plate. The lowest part of our planet is 7 miles (11 km) below sea level. Earthquakes are commonplace, as is volcanic activity, and a huge ring of volcanoes encircles the Pacific Ocean. Plate tectonics renews the surface crustal material, mainly at the 37,000-mile (60,000-km) long mid-oceanic ridges. As a result of this constant renewal, there are only about 170 known impact craters on the Earth's surface, in contrast to the huge numbers that can be seen on neighboring Venus, Mercury, and the Moon. It would appear that for every new crater that is produced, an old one is destroyed.

Moving continents

The movement of the tectonic plates also moves the six continents around. Five hundred million years ago, the south spin pole was in Africa, near the center of the Sahara Desert. Antarctica was then close to the equator. Typically the plates are moving at a few inches a year. This may sound slow, but it means that the whole Atlantic Ocean has been produced in 250 million years, which represents less than 6 percent of the lifetime of the planet. This continual movement and activity has ensured that hardly any evidence of the earliest chapters of Earth's history remains. The first 60 percent of our planet's evolution remains very much a mystery.

◀ Alaskan heat wave

This nearly cloud-free view of Alaska was captured by NASA's Terra satellite on June 17, 2013, when the region was experiencing a heat wave. Temperatures reached record highs of up to 96°F (36°C). On most days, Alaska is obscured from view by banks of cloud that relentlessly stream over the region, and some parts of the south coast average more than 340 cloudy days per year. On this rare clear day, the rich tapestry of forests, rivers, and mountain ranges was exposed to view. The ice-capped Alaska Range mountains stretch across the south of region.

▶ Mir Space Station

The Russian Space Station Mir was photographed from the Space Shuttle Atlantis on July 4, 1995. A week earlier, the shuttle had docked with the station. Mir operated in low orbit around Earth from 1986 to 2001, with an uninterrupted human presence in space of 3,644 days. The record for the longest single human spaceflight is held by cosmonaut Valeri Polyakov, who spent 437 days on Mir from January 8, 1994 to March 22, 1995.

► **Algal bloom**

A huge bloom of phytoplankton can be seen in the oceans off the coast of Argentina in this image taken on December 21, 2010 by NASA's Aqua satellite. Aqua has orbited Earth since 2002, studying evaporation, precipitation, and cycling of water. Phytoplankton are essential to all life in the oceans, and account for about 50 percent of the world's photosynthesis, capturing the energy of the Sun. Recent studies have shown levels of phytoplankton in surface waters declining by around 1 percent per year, which is having a serious impact on the entire marine ecosystem.

▲ **Grand Canyon**

Carved by the Colorado River, the Grand Canyon has vertical drops of 5,000 feet (1,500 meters). The rock strata on the steep walls form a geological record ranging from the Precambrian Era 1.7 billion years ago to the Paleozoic Era 250 million years ago.

Water on Earth

The extensive oceans on Earth are at least 3.5 billion years old. The continents move around, but the relative areas of land and ocean have changed little during this period. The continents are made up of low-density rock that is floating on top of higher-density mantle rock. The atmospheric contents and the water are continually being recycled. Water evaporates, rises, and condenses around dust particles to form clouds.

The circulation patterns in the atmosphere are driven by sunlight. Evaporation occurs mainly near the equator; trade winds carry this water to other regions before releasing it. Rainfall thus varies greatly as a function of time and place, with the result that some regions of the continents are lush rain forests, and others sunbaked, dusty deserts. From space, the land areas of Earth appear either dark green or in shades of yellow-brown. Forests cover about 30 percent of the land surface, and deserts 20 percent. The global water cycle carries water from the oceans up into the clouds; these are then blown to the tops of mountains, where the water falls as rain or snow. Running water and expanding ice erode surface features and the rock is slowly ground down, producing sediments. Some of these processes take place relatively quickly, and the Grand Canyon in Arizona has been carved by the Colorado River in only 10 million years.

Heat storage

About 37 percent of the solar radiation that reaches Earth is reflected back into space. The percentage that is reflected varies depending on the cloud coverage—on average, about 50 percent of the planet is covered in cloud at any one time. Greenhouse gases in the atmosphere, such as carbon dioxide and methane, absorb some of the radiation reflected by the surface, and the warmed gases re-radiate the heat in all directions, including back towards the surface. Temperatures at the surface are about 55°F (30°C) warmer than they would be without this process, meaning that there would be little liquid water without it. A great deal of the heat at the surface is stored in the oceans, even though the average water temperature is only 39°F (4°C), compared to an average air temperature at the surface of 59°F (15°C). This storage helps to stabilize the planetary temperature.

Earth's spin axis is inclined at an angle of 23.5 degrees. It is this angle that produces the strong seasonal variations in temperature and daylight length. Differential heating of different parts of the globe set up large-scale circulation currents in the atmosphere and the oceans, transferring heat and moisture across the planet. The variations in climate caused by these circulation patterns have huge effects on plant growth cycles, rainfall, and animal migration.

Earth's core

The center of Earth is just as inaccessible to us as the surface of the Sun. When descending into a deep mine we see that the temperature increases as we go down. Whenever we are near a volcano, we realize that the molten rock is only a few miles

down and that spasmodically it breaks through the surface and flows out. Physical modeling indicates that the density and pressure of the planet increase with depth; in fact, the pressure at the center of our planet is about four million times greater than the pressure at the surface. The temperature is approximately 11,000°F (6,000°C).

Unknown composition

We are not sure about the exact composition of Earth's interior. Knowing the mass and volume of our planet, we can work out that the average density is about 5.52 times that of water. The continental surface rocks have a density only 2.9 times that of water, indicating that the deep interior must have the high relative density of about nine. The interior probably contains a lot of iron and nickel. Meteorites provide the best clue to

▼ Sahara Desert

High clouds form over the Sahara Desert. Earth's hot deserts are found across the subtropical continents. Air that warmed and ascended near the equator falls again over this region, creating high pressure and dry, mostly cloudless conditions. Intense heating of the ground during summer may occasionally create rainclouds, but any rain that falls is likely to evaporate before it reaches the surface.

the original components of Earth, and suggest that the proportion of the elements in Earth is (by mass) 35 percent iron, 29.5 percent oxygen, 15 percent silicon, 12.5 percent magnesium, 3 percent nickel, 1.5 percent sulfur, 1 percent calcium, 1 percent aluminum, and 1.5 percent other material.

Scientists investigate the core by carefully noting the way the planet shakes during earthquakes. Using land-based seismometers, geophysicists measure the speeds of the quake-generated pressure waves at different depths; the speeds are a function of the density and strength of the material that the waves pass through. It seems that the composition of Earth changes from rock to an iron-nickel alloy at a depth of about 1,800 miles (2,900 km). Below this depth is an iron-nickel core. At a depth of 3,200 miles (5,150 km), this core becomes solid.

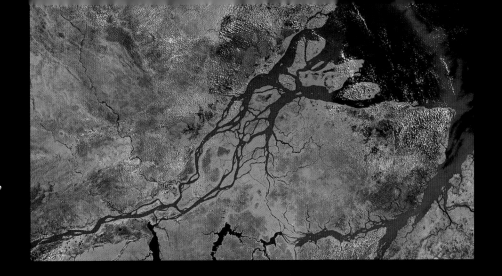

▲ **Amazon Delta**

Year-round rainfall produces areas of lush forest around the equator. The Amazon Rain Forest in South America is the largest. The Amazon River flows through the forest and discharges into the Atlantic Ocean at a huge delta. It discharges 7.4 million cubic feet (209,000 cubic meters) of water per second—more than the next seven largest rivers combined.

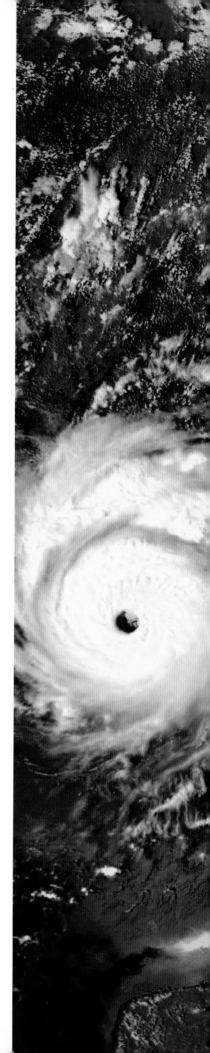

▲ Blue skies

The atmosphere appears blue in this image taken from the International Space Station. This effect is caused by air molecules scattering blue light, which is why the sky appears blue from the ground. The atmosphere also reflects light, which is why the lower part of the Moon appears to fade away in this picture.

► Hurricane Andrew

This time-series of three images shows the position of Hurricane Andrew on August 23, 24, and 25, 1992, as it moved east to west over south Florida. Hurricanes, or tropical cyclones, are weather systems that form over tropical oceans. Rapid evaporation from the warm water creates storm clouds, which are set into a spiralling motion by the Coriolis effect, caused by the rotation of Earth. In the northern hemisphere, tropical cyclones spiral counterclockwise, while in the southern hemisphere, they spiral clockwise.

Earth's atmosphere

The pressure and density of Earth's atmosphere, like those of all planets, decrease as a function of height. On Earth these quantities go down by a factor of ten for every 10.25 miles (16.5 km) of height. If the density were constant, the atmosphere would be only 5.3 miles (8.5 km) thick. Forty-one percent of the mass of the atmosphere is contained in the lowest 3 miles (5 km). Twenty-one percent of the molecules in the Earth's atmosphere are oxygen and 78 percent are nitrogen.

About 30 miles (50 km) above the surface is an ozone layer. Ozone is very important because it absorbs ultraviolet radiation and thus makes the surface safer for living things. The existence of oxygen, which is a highly reactive gas that combines chemically with many other elements, is due to the activity of plant life, which converts carbon dioxide into oxygen by photosynthesis. Plant life accounts for the unusually small traces of carbon dioxide, too.

All weather on Earth occurs in the lowest layer of the atmosphere, called the troposphere. Above the poles, the troposphere is about 5 miles (8 km) thick. It is thickest at the equator, where it extends about 10 miles (16 km) up. Above the troposphere is the stratosphere, which is more stable. Conditions in the stratosphere are roughly equal all around the planet. Temperatures drop as you ascend through the troposphere, but rise again through the stratosphere. Above the stratosphere, temperatures fall again in the mesosphere, only to rise once more in the outer thermosphere.

Magnetosphere

Earth has the strongest magnetic field of the rocky planets. This is produced by circulating currents in the molten metallic core. Earth's magnetic field extends for tens of thousands of miles in all directions. The magnetosphere deflects the stream of charged particles emanating from the Sun known as the solar wind, and life on the surface of the planet is protected from dangerous radiation. Some of these charged particles become trapped in two donut-shaped areas called the Van Allen belts. Here, energetic particles spiral back and forth from pole to pole at speeds of thousands of miles per second.

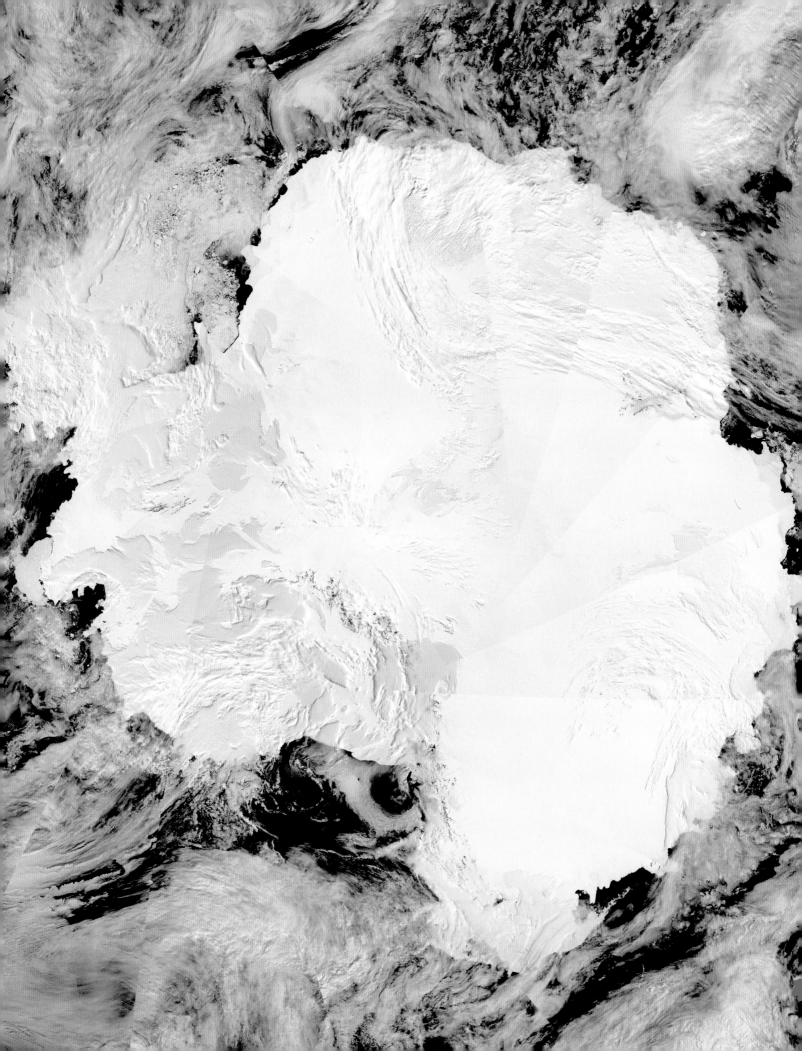

The Antarctic ice-sheet is by far the largest glacier on Earth. It covers almost the whole of the continent and contains more than 70 percent of the planet's fresh water. If the ice sheet were to melt, the oceans would rise by 230 feet (70 meters). Some parts of the ice sheet are rich in meteorites, which fall onto the ice and become buried in it.

Rocky crust

The crust has an average thickness of about 18 miles (30 km) below the continents and 6 miles (10 km) below the oceans. The crust is continually forming new rock as existing rocks are changed and eroded.

The continental crust contains a wide variety of rocks, some of which are 4 billion years old. Sedimentary rocks, such as sandstone, coal, and limestone, are formed from the compression of layers of sediment eroded from the continents, which may be lifted out of the sea to form mountain ranges. Igneous rocks, such as granite and gabbro, form from lava that erupts to the surface in volcanoes. Metamorphic rock, such as marble and slate, form from sedimentary rock or igneous rock that has been subjected to high temperatures and pressures.

The oceanic crust is largely formed of dense basaltic lavas, which are no more than 200 million years old. Oceanic crust is continually forming from mantle that emerges from underneath the crust to form ridges. These can grow large enough to emerge from the ocean as chains of volcanic islands such as the Hawaiian Islands in the Pacific Ocean.

Ice ages

Earth's climate changes over time. At intervals of about 100,000 years, during colder periods called ice ages, vast sheets of ice some 1.8 miles (3 km) thick can be produced, covering much of the Northern Hemisphere. The average temperature drops by about 5.5°F (3°C), the level of the oceans goes down, and the circulation patterns in the atmosphere change. Ice moving south can carve out huge basins. The Great Lakes in North America are a typical example, and now contain about 20 percent of Earth's fresh water.

Ice ages are caused by changes in the shape of Earth's orbit and in the tilt of its spin axis. The last ice age ended 10,000 years ago, and we are now entering a warm phase.

decades, global temperatures have risen by about 2°F (1°C), and the annually deposited layer of arctic ice has decreased in thickness by 40 percent since 1960. This is due mostly to anthropogenic climate change—an increase in temperature caused by the release of the greenhouse gas carbon dioxide into the atmosphere as fossil fuels are burned for energy.

Snowball Earth

Some scientists believe that the whole of the planet may have been covered in ice at least once in its history. The "Snowball Earth" hypothesis explains evidence of past glaciation at tropical latitudes. Even if the planet was not entirely ice-bound, it is clear that Earth experienced an extreme ice age from 750 million years ago to 580 million years ago. At other times, there has been no ice at all. The last "hothouse" period occurred 55 million years ago, when the Arctic Ocean warmed to a sub-tropical 74°F (23°C).

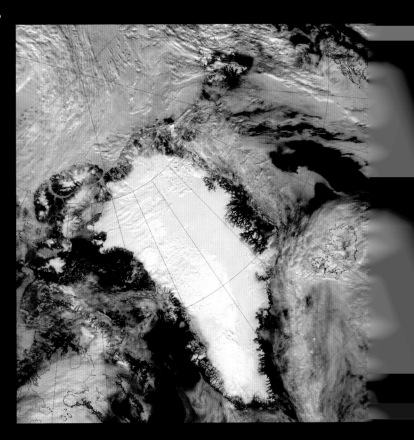

▲ **Greenland Ice-sheet**

Eighty-five percent of Greenland is covered in a single ice-sheet, which contains 10 percent of Earth's fresh water. Due to the albedo effect—the high level of reflectivity of light-colored surfaces—much of the Sun's radiation is reflected by the ice-sheet, keeping summer temperatures low. Satellites monitor changes in the surface height of the ice using lasers, looking out for signs that it may be shrinking due to climate change.

▶ **Earth at night**

This composite of satellite images shows how Earth is lit up at night by human activity. The most lit-up areas indicate dense human populations combined with high levels of urbanization. Humans are rapidly urbanizing, and the urban population of Africa is projected to rise from one-third in 2010 to one-half in 2030.

Living planet

Earth is the only place we know of with life. Research has revealed that the evolution of life was a very slow process. But the fossil record indicates that as soon as the surface of our planet became both relatively stable and cool enough to be wet, primitive life broke out. It is estimated that this occurred about 3.8 billion years ago. The asteroidal impact rate had slowed down by then, and the crust had become thick enough for the early continents to last for a reasonable time before being dragged back into the mantle and destroyed. The condensed water had formed huge seas and oceans. At that time the Moon orbited closer to Earth, the days were short, and the seas were affected by very high tides.

▲ **Deforestation**

This satellite image shows the extent of deforestation on the island of Hispaniola in the Caribbean. To the right, lush green forests still cover large areas of the Dominican Republic. To the left is Haiti, where forests have been destroyed by loggers. In 1920, 60 percent of Haiti was forested. Today the figure is less than 2 percent.

What is life?

Defining life is not straightforward. To be classed as alive, an organism must have an active internal chemistry, which it powers by taking in energy. It must also be able to grow and reproduce itself. For the first 3 billion years, life consisted solely of simple organisms such as bacteria and archaea, and even today, the total biomass of single-celled life may be more than that of multicellular plants and animals, most of it too small to be seen by the naked eye. The first large, complex multicellular life forms appeared during the Cambrian Explosion, which started 580 million years ago. Levels of oxygen in the atmosphere increased due to photosynthesis by

plants and algae. The ozone layer formed in the atmosphere, and this allowed life to thrive on land without being damaged by ultraviolet radiation. Over the course of a few tens of millions of years, life on Earth diversified.

Human impact

Modern humans first appeared in Africa around 250,000 years ago. Since then, we have colonized every continent, including Antarctica. Following the development of agriculture in Mesopotamia around 12,000 years ago, the increase in human impact on the surface of the planet has been dramatic, and today 80 percent of the land bears the mark of human activity. With industrialization, human population levels have grown from 1 billion in 1800 to 7 billion today.

With the growing human population have come growing pressures on resources. Deforestation has reduced the proportion of land covered in forest to 30 percent. This, combined with the release of carbon stored in fossil fuels, has contributed to an increase in the level of greenhouse gases in the atmosphere. Scientists predict that over the coming decades, global temperatures are set to continue rising.

MOON

Our Moon is odd. Even though its mass is 81 times less than Earth's, it is still the fifth-largest moon in the solar system. No other terrestrial planet has a moon like it. Why Earth is blessed with such a large Moon is a great mystery. It is the brightest object in the night sky and exerts a huge influence on our planet. Its changing phases and fixed orbital period define the month. The interval between first quarter and second quarter, being seven days, probably defined our seven-day week. And the gravitational pull of the Moon on the waters of Earth produces most of the tidal changes in sea level. The Moon is dry, airless, and dead. Its mean density is 3.3 times that of water, rather similar to the density of the mantle of Earth. So it is like Earth, but without our planet's core of iron and nickel. The relative proportions of the isotopes of oxygen are similar, suggesting that the Moon and Earth were formed out of the same mixture of rocky materials, and at the same distance from the Sun in the original preplanetary cloud of gas and dust.

▼ View from the Moon

This iconic image of Earth was photographed from lunar orbit during the Apollo 8 mission, on Christmas Eve 1968. Earth's beautiful blue color is caused by the scattering of the shorter (bluer) wavelengths of light as they encounter molecules in the planet's atmosphere. By contrast, the lifeless Moon reveals only its dry, rocky, atmosphere-free surface.

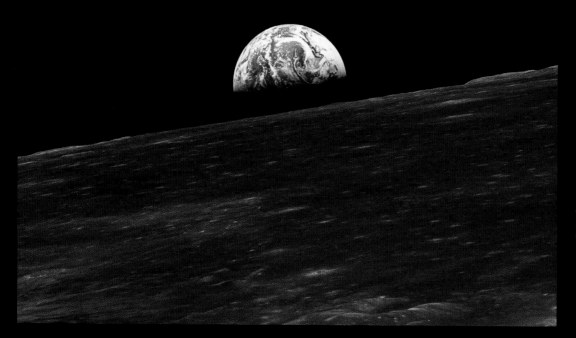

▶ Sea and mountain

Seen from Earth, the Moon's near side has noticeably dark regions—these are huge, lava-filled crater basins called maria (Latin for "seas"). The whole surface was once covered in craters, most of which were produced about 4 billion years ago during a time of massive bombardment. At that time, the Moon was volcanically active. Lava rose to the surface through cracks and fissures, filling the lower parts of the largest craters to form the maria. Lighter areas are higher land. The plains reflect only about 4 percent of the sunlight that hits them, whereas the mountains reflect 11 percent.

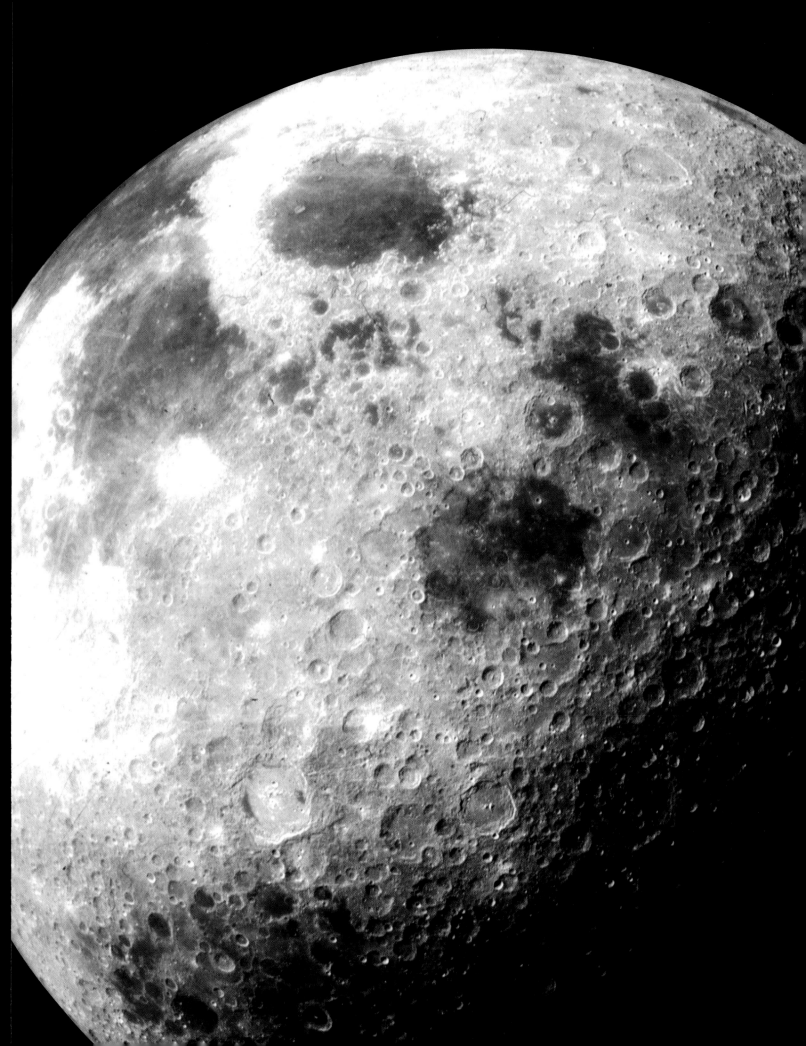

Origins of the Moon

Where did the Moon come from? No one knows exactly how it was formed, but there have been two notable theories. A hundred years ago the favorite theory was known as spin bifurcation. As the early Earth condensed, it spun faster and faster because of the conservation of momentum. The spinning caused Earth to become progressively more pear-shaped. Instability then supposedly caused this "pear" to break in two, and the larger part became Earth and the smaller piece became the Moon. Some scientists even suggested that the scar produced by this fracture is still visible on Earth as the Pacific Ocean basin. However, there are many problems with this idea. The present Earth-Moon system does not have enough momentum to suggest that the initial proto-Earth's spin became unstable. Also, when this fluid bifurcation takes place in other situations, the mass ratio of the two final bodies is usually about 12 to 1 and not the 81 to 1 seen with the Earth and Moon.

The impact theory

More recently, most astronomers have tended to agree with the giant-impact theory, which hypothesizes that the process was set in motion about 4.5 billion years ago when a massive asteroid, about the size of Mars, hit the young Earth and knocked off a section of the outer mantle. The ejected material formed a massive cloud of gas, dust, and rock. Heat was radiated away and the cloud quickly began to cool. Most of the ejected material went into orbit around Earth, forming a clumpy, dense, donut-shaped ring. Rocks grew by mutual collisions until a single body dominated the ring, sweeping up the remaining material, and the Moon was born.

There is supporting evidence for the giant-impact theory. Earth's spin and the Moon's orbit have similar orientations. Lunar samples reveal that the Moon's surface was once molten. Certainly, the giant-impact theory is consistent with the theories about the formation of our Solar System.

The Moon's original orbit was close to Earth, but as time passed, both tidal friction and momentum exchange made the new satellite recede. As it did so, Earth's rotation rate slowed down and its day gradually increased in length. But why did only one large asteroid hit Earth in those early times? Why was only one moon formed? And if the process worked so well for our planet, why did similar collisions not produce moons for Mercury, Venus, and Mars? These questions have yet to be answered.

▶ **Eclipse**

As seen from Earth, a solar eclipse occurs when the Moon passes between the Sun and Earth, and the Moon fully or partially blocks ("occults") the Sun. This can happen only at new moon, when the Sun and the Moon are in conjunction as seen from Earth—in an alignment known as syzygy. In a total eclipse, the disk of the Sun is fully obscured by the Moon.

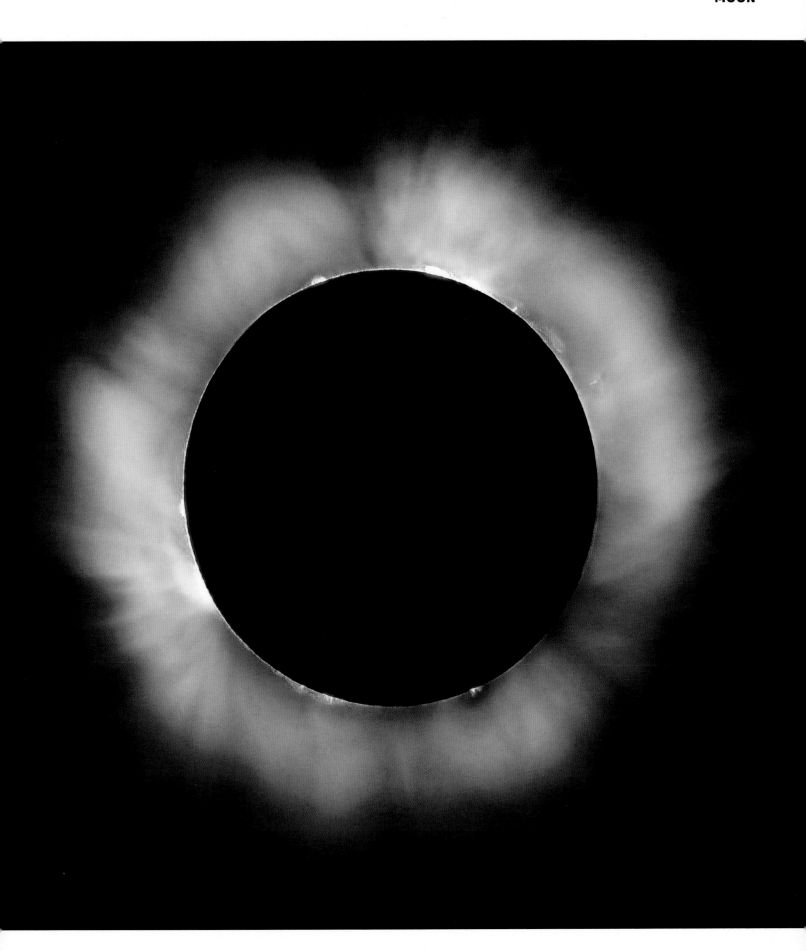

Moon craters

The old highland regions of the Moon are pockmarked with craters. These were produced mainly in the first 750 million years of the Moon's life, when large numbers of small asteroids hit its surface and cratered and cracked the crust. As time passed, the asteroid bombardment became less intense. The radioactive decay of elements in the lunar rocks slowly heated up the outer regions, and between 3.5 and 3.2 billion years ago there was a period of extensive volcanic activity, when lava oozed up through the cracks and filled the low-lying basins. These dark, lava-filled regions are younger than the highland regions and the crater density is lower. The samples returned from the Moon have enabled us to calculate the ages of some features precisely.

Classifying the craters

Craters come in three types. Those smaller than 7 miles (11 km) wide are bowl-shaped. Their depth is about a third of their diameter, just like bomb craters on Earth. With craters 7 to 90 miles (11 to 150 km) wide, the initial impact crater tends to be altered because of the low strength of the lunar crust. The crater sides have suffered landslips, and the rebound of the underlying lunar material has produced mountains in the center. Craters larger than 90 miles (150 km) wide have suffered many landslides and usually have a series of circular mountain rings around them. Here, the initial crater was so deep that lava from under the lunar crust surged to the surface and filled in the low-lying regions. Even though the incoming asteroids hit the surface at a range of angles, the fact that the craters are about twenty times larger than the asteroid means that most craters are circular.

Two of the largest craters on the moon are Mare Orientale, 560 miles (900 km) wide, and South Pole Aitken, 1,550 miles (2,500 km) wide. The size distribution of craters is such that for every crater larger than 60 miles (100 km) wide there are 100 larger than 6 miles (10 km) wide, and 10,000 larger than 0.6 mile (1 km) wide.

The far side

The spin of the Moon is such that the same face is always pointing toward Earth. If we want to see what is on the other side we have to send a spacecraft to look. When this was

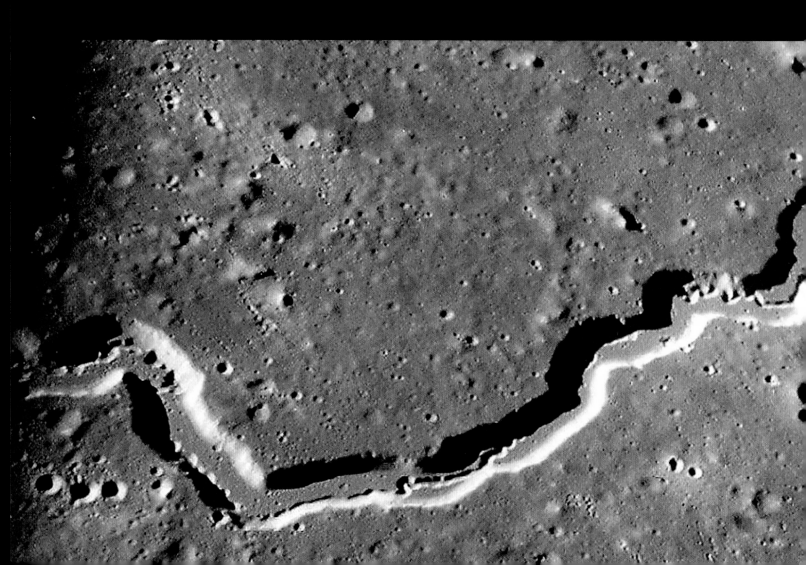

done in the early 1960s, one of the major surprises was that much less lava had filled the low-lying regions of the far side. This was because the rocky crust on the far side is about 20 miles (30 km) thicker than on the near side.

Inhospitable surface

The surface of the Moon is covered with a soil-like material called regolith. This porous blanket of rubble has been produced by more than 4 billions years of meteoritic bombardment. The rubble is yards thick, and particle sizes range from fine-grain dust to huge boulders. Underlying the regolith is fractured rock; only below a depth of 15 miles (25 km) is the rock undamaged.

When the Sun is overhead at the Moon's equator, the temperature is about 230°F (110°C), higher than that of boiling water. Two weeks later, when that region has spun around and noon is replaced by midnight, the temperature has dropped to –274°F (–170°C), cold enough to freeze air. Clearly, living on the Moon would not be easy. The fact that there seems to be no oxygen and very little water does not favor potential colonization.

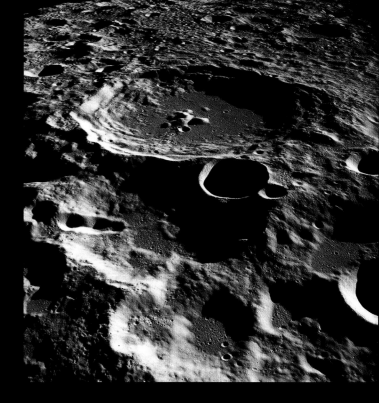

▲ Crater shadows

The Moon's larger craters—those between 6 and 90 miles (10 and 150 km) in diameter—have outer walls that have slumped into the crater pit. There is often a central mountainous peak produced by the recoil of the underlying stressed rocks. These craters were photographed just after dawn, when the low eastern Sun casts long shadows that emphasize the difference in height between the floor and the rim.

◀ Moon valley

Schroter's Valley is a huge valley or rille (a long, narrow depression that resembles a channel) on the near side of the Moon. It has a maximum width of about 6.2 miles (10 km) and narrows to less than 0.6 miles (1 km) near its terminus. The rille probably formed when a lava tube collapsed billions of years ago. Similar rilles have been observed on Mars.

Destination Moon

In 1961, President John F. Kennedy set a national goal of "landing a man on the Moon and returning him safely to the Earth" by the end of the 1960s. The missions that took 12 men to the Moon between 1969 and 1972 were part of NASA's Apollo program.

The enterprise rested on three radical ideas. First, the Apollo spacecraft would have sections that could be discarded en route to reduce the overall mass. Second, instead of heading straight from Earth to the Moon, two intermediate orbits around Earth and the Moon were used to give the crew time to check their equipment and prepare for landing. Third, the crew divided when they reached lunar orbit. Two descended to the Moon, while the third remained in orbit.

A three-stage Saturn V rocket would lift the Apollo craft into low Earth orbit. Its upper stage would then blast the craft on its journey to the Moon.

Blast off

Apollo 11, the first manned mission to the Moon, blasted off from Cape Kennedy (now Cape Canaveral) in Florida on July 16, 1969, and entered orbit around the Moon three days later. Astronauts Neil Armstrong and Edwin "Buzz" Aldrin then climbed aboard the Lunar module "Eagle" for the descent to the surface, leaving Michael Collins aboard the command and services module "Columbia" in lunar orbit. "Eagle" headed toward the Mare Tranquillitatis, a huge impact crater filled with basaltic lava. The site had been chosen carefully. It was fairly flat, which was important not only for landing but for taking off. The landing time had also been calculated. It had to take into account the temperature of the lunar surface, which varies drastically during a lunar day. As he closed in on the landing site in the "Eagle," Buzz Aldrin could see that it was strewn with boulders. He took manual control of the craft and headed for a smoother spot. With only 20 seconds of fuel left, "Eagle" touched down.

▲ **One giant leap**

Neil Armstrong steps down the ladder of the "Eagle" and sets foot on the surface of the Moon. Moments later he uttered his famous words, "That's one small step for [a] man; one giant leap for mankind." There are few images of Armstrong himself on the Moon as he was responsible for holding the camera.

◄ **Lunar experiments**

Here, Buzz Aldrin sets up a solar wind experiment on the lunar surface. Soil and rock samples were also collected in sample bags. Armstrong and Aldrin then experimented with methods of moving around in low gravity, including kangaroo-style hops.

"Eagle" landed about 4 miles (6 km) away from its target spot. Although scheduled to sleep, the astronauts were given clearance to prepare for their first surface excursion.

Walking on the Moon

Six hours and 21 minutes after landing, with life support systems on their backs, the astronauts depressurized the cabin and opened the module's hatch. Watched by millions of television viewers on Earth, at 2:56 a.m. on July 20, Neil Armstrong backed through the hatch and moved down the ladder on the outside of the Lunar Module to become the first human to step onto a world other than Earth. Armstrong and Aldrin stayed on the surface for about

21 hours, completing one moonwalk. They planted the U.S. flag and set up a number of experiments, collected rock samples, and spoke to President Richard Nixon by telephone. After making a successful ascent from the surface and a rendezvous with Michael Collins in "Columbia", they returned safely to Earth three days later.

After Apollo 11

Five further missions landed on the Moon. All six landing sites were on the near side as direct communication between the lunar surface and Earth is not possible from the far side. The landing sites were also close to the Moon's equator, as the regions around the lunar poles are too cold.

▲ Footprint of fame

One of the most famous photographs of all time is this image of Neil Armstrong's footprint on the Moon. Given the Moon's lack of atmosphere, this footprint could remain there intact for longer than the human race survives.

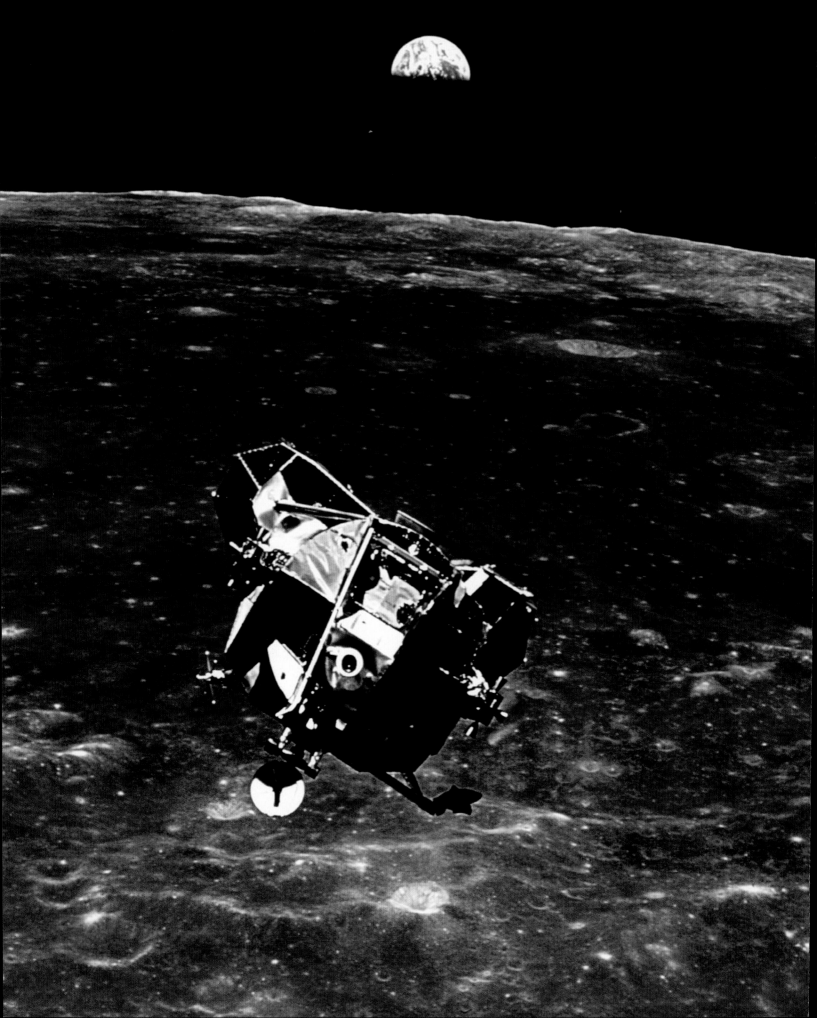

Here, the Apollo 11 Lunar Module carrying Neil
Armstrong and Buzz Aldrin approaches Michael Collins
inside the Command and Services Module, as Collins
films the view. Delicate maneuvering allowed the two
craft to dock. Since the module would never have to fly
in an atmosphere, the designers felt free to choose a
functional shape. However, the cabin was so cramped
the astronauts had to stand up during the flight.

Working on the Moon

Once they had arrived on the Moon the astronauts had very
little time in which to conduct experiments and explore.
The Apollo 11 team had two-and-a-half hours. By Apollo 14,
the moonwalking time had been increased to nine-and-a-half
hours, while on the final mission, Apollo 17, exploration time
was 22 hours. The first three successful missions (11, 12, and
14) conducted explorations on foot, and could not stray far
from the landing sites. Later missions, however, had the
advantage of a Lunar Roving Vehicle.

After leaving the pressurized safety of the Lunar Module,
the astronauts depended on their spacesuits. Though
cumbersome, the suits protected them from X-rays, ultraviolet
radiation, and from the vacuum and cold of space. Its fabric
was strong enough not to tear against sharp rocks on the
surface. The life support system strapped to the astronauts'
backs was bulky and made getting in and out of the Lunar
Module a challenge, but this system kept the astronauts cool
and supplied with oxygen. The suit's flexibility was also
important: its rubber joints were stretchy, and its gloves were
moulded individually to allow for a maximum sense of touch.
Walking was made easy by the Moon itself. With only one-
sixth of Earth's gravity, hills were easy to climb.

Astronauts conducted moonwalks for two reasons. They
collected rock and soil samples, and they set up instruments
to record data after the mission had ended. On the Apollo 14
mission, Alan Shepard and Edgar Mitchell spent their first
four hours of surface activity setting up instruments. Their
second excursion was a tough climb to the rim of Cone
Crater, about 4,600 ft (1,400 m) from the Lunar Module.
Pushing a wheeled cart, which they filled with rock samples,
theirs was the Apollo program's longest moonwalk.

Lunar driving

The last three trips to the Moon, Apollos 15, 16, and 17, each
took a Lunar Roving Vehicle. The rover traveled to the
Moon folded up on the side of the Lunar Module, and was
lowered to the surface on a pulley. The rover allowed the
astronauts to explore an area much larger than they could
on foot. It was designed to travel a distance of 56 miles (96
km), at a speed of up to 11.5 mph (18.5 kph). However, the
astronauts were told not to drive more than 6 miles (10 km)
from their Lunar Module. If the rover failed for any reason,
they would be able to walk back before their oxygen ran out.
The furthest traveled on one trip was 4.7 miles (7.6 km), on
Apollo 17 in 1972.

The astronauts experienced two problems while roving
across the Moon. The glare from the Sun made it difficult to
choose their route, and made obstacles such as boulders
almost impossible to see because there were no shadows. The
second was the Moon's dust, which was electrostatically
charged and clung to all surfaces, so the rover's wheel guards
were vital for keeping the equipment clean.

▲ Roving over the moon

At 10 ft (3 m) long and 6 ft (2 m) wide, the Lunar Roving Vehicle was about
the size of a Volkswagen Beetle car. Made of aluminum tubing it was light
and strong, with four shock-absorbing, wire-mesh wheels equipped with
titanium treads. It was powered by battery, weighed about 460 lb (210 kg),
and could carry two astronauts and their life support systems, scientific
equipment, and camera gear—a load three times its own weight.

MARS

Mars, the red planet, has only 11 percent the mass of Earth and is just over half the size. Its mass is easily calculated because the planet has two small, asteroidlike satellites that it probably captured in its infancy. Mars is the outermost of the four rocky planets and is about 50 percent farther away from the Sun than Earth, so it is much colder. The average temperature is −81°F (−63°C). It never rains, but frost in the polar regions creates white polar caps.

Mars has an orbit that is much more eccentric than that of Earth; when it is at its closest to the Sun, it receives 45 percent more radiation than when it is at its farthest distance from the Sun. A day on Mars is only twenty four minutes longer than an Earth day. The spin axis tilt is 25.2 degrees, so Mars has marked seasonal variations in weather.

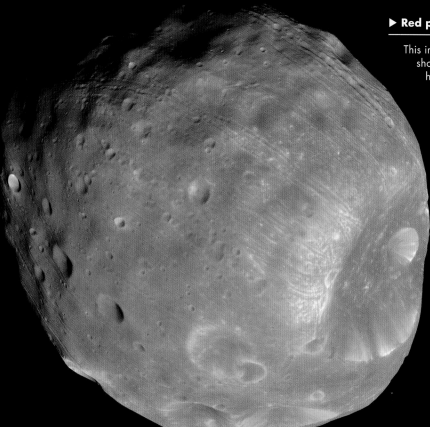

▶ Red planet

This image taken by the Viking orbiter clearly shows the northern polar cap, the northern hemisphere of low volcanic plains, and the higher, older ground to the south. In the southern hemisphere, the Valles Marineris can be clearly seen running west to east below the equator. This is an enormous system of canyons formed by a splitting of the surface when the planet was young. Also visible to the left are the vast volcanoes of the Tharsis region.

◀ Small moons

Mars has two tiny moons, called Phobos (shown here) and Deimos. Many astronomers believe that these small, irregular-shaped bodies are asteroids that were trapped by Mars's gravity. Phobos is 16.6 miles (26.8 km) long, while Deimos is just 9 miles (15 km) in length.

The Martian surface

Mars has always fascinated observers. It is outside Earth's orbit, so it is easier to see than Venus and Mercury. In the early days of the 20th century, astronomers thought that they had seen long, dark, linear features on the surface. Active imaginations quickly converted these into canal systems, and theories of civilizations irrigating the tropical regions with polar meltwater abounded. These ideas evaporated when the first spacecraft arrived.

Orbiters mapped the surface in detail, and landers, such as Viking 1 in 1976, showed a red, rock-strewn, dry, dusty, and windswept desert freezing under a pink sky. Small roving vehicles, such as the 1997 Mars Sojourner, analyzed some of the rocks. These were volcanic lavas with an enhanced

basaltic composition, being rich in volatiles such as sodium, potassium, and sulfur. Maps from the satellites showed that the two hemispheres of the planet were dissimilar. The southern hemisphere is 0.6 to 2.5 miles (1 to 4 km) higher than the mean level. It is old, and is profusely pocked with impact craters. The northern hemisphere is low-lying, with the exceptions of a few huge volcanoes, but the reason for this disparity is still unknown.

The orbiters also revealed subtle differences between Martian impact craters and lunar ones. On Mars the effect of wind erosion is obvious. Martian craters resemble the craters formed when pebbles are dropped into mud, and there is every expectation that Mars has some water hidden below the surface. Cratering events melt this water, and muddy

slurry flows from the impact site, but it quickly freezes before it has traveled far. The amount of water, and its depth, is likely to vary considerably depending on the latitude. There is every possibility that Mars had, in the distant past, enough water to cover the planet (if completely spherical) to a depth of half a mile (0.9 km). But the water has long since been frozen or evaporated.

When it comes to the question of life on Mars, the main problem is not that the planet is too cold, but that it is too small. Life requires an atmosphere that can produce an efficient greenhouse effect, and one that can hold water, have clouds, and transport water from seas and lakes to mountaintops. Mars's core cooled down long ago, and as a

The Martian atmosphere

Mars has a thin atmosphere that is 95.4 percent carbon dioxide. The rest is 2.7 percent nitrogen, and 1.6 percent argon. The ground pressure is a meager 0.65 percent that at the surface of Earth. From the Martian surface the atmosphere looks pink because of the haze of iron oxide (rust) particles suspended in it. Sometimes, thin, white clouds of frozen carbon dioxide and water ice form above the high mountains. Mars can be very cloudy, but these clouds are usually dust, caused by storms that follow a seasonal pattern and envelop the entire planet in a dust cloud. In fact, observation by orbiters shows that Mars experiences daily weather changes. On a typical summer's day

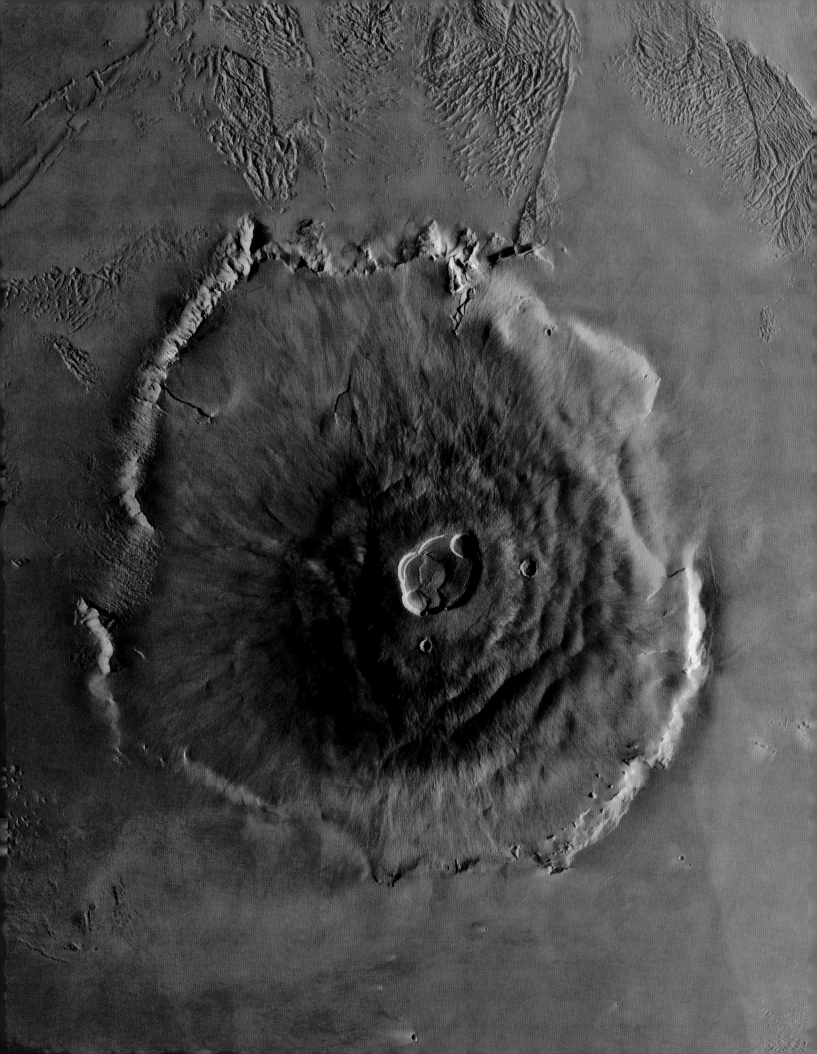

Volcanic Mars

Volcanism has played a significant part in Mars's geological evolution, and the planet may still be volcanically active today. Volcanic features include the extensive lava flows and lava plains that cover huge portions of the surface, and the largest known volcanoes in the Solar System. These features were formed about 3 billion years ago when the planet was still young. In contrast to Earth, where tectonic plates move over rising magma and new volcanoes form in different spots as the plates shift position, the volcanic regions of Mars remained fixed in place on the same part of the planet's surface, so volcanoes at specific spots continued to grow with each successive eruption. The crust of Mars can support such huge mountains because it is colder, thicker, and stronger than Earth's, and the gravitational field is weaker.

Olympus Mons

Olympus Mons, the youngest of the giant volcanoes, lies is in the Tharsis region—a huge, domed plateau on the planet's equator. The volcano is more than 15 miles (24 km) high—more than three times the height of Mount Everest on Earth—and the base diameter is more than 370 miles (600 km) wide. It is similar in appearance to the Hawaiian shield volcanoes, encircled by wide terraces of solidified lava, but has a volume that is 100 times greater. The complex caldera at the top is 32 miles (52 km) wide. The different areas of the caldera floor date from different periods of volcanic activity. The caldera is encircled by a huge scarp, up to 3.7 miles (6 km) high. Vast plains, known as aureoles, extend outwards from the summit, like the petals of a flower, and end in a giant surrounding scarp of up to 5 miles (8 km)—a unique feature of Martian shield volcanoes. The whole volcano covers an area about the size of Arizona. Farther out, lava covers the vast Tharsis plateau.

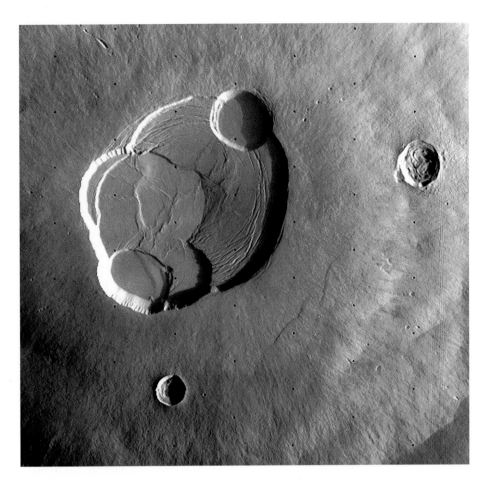

Volcano family

The next three largest volcanos in the region—Ascraeus Mons, Pavonis Mons, and Arsia Mons—line the top of the plateau.

Arsia Mons is second only to the mighty Olympus Mons in terms of volume. Its summit rises 5.6 miles (9 km) above the surrounding plain.

Volcanoes and ice

Large amounts of water ice may be present in the Martian subsurface. When lava comes into contact with surface ice, vast amounts of liquid water and mud may form massive debris flows (lahars). Some channels in volcanic areas, such as Hrad Valles near Elysium Mons, may have been carved or modified by lahars. Lava flowing over water-saturated ground can cause the water to erupt in an explosion of steam, producing volcano-like landforms called pseudocraters.

▲ Giant caldera

In this bird's-eye view of the nested caldera on the summit of Olympus Mons, five roughly circular areas of caldera floor can be seen. The youngest calderas form circular collapse craters. Older calderas appear as semicircular segments because they are transected by the younger calderas.

◄ Mighty Mons

An overhead image of the Olympus Mons aureole taken by the Viking Orbiter gives it a deceptively flat appearance. Because of its sheer size and its shallow, shield-volcano slopes, an observer standing on the Martian surface would be unable to view the entire profile of the volcano, even from a great distance.

4 miles (7 km) deep. In comparison, Arizona's Grand Canyon is about one-tenth as long and one-fifth as deep.

The Valles Marineris runs roughly west to east just below Mars's equator, and seems to be a giant rift valley, which means that parallel faults split the crust and the land between them fell. This is in contrast to the Grand Canyon, which formed largely through water erosion. Landslides have modified the sides of the Marineris system, increase the width and depth of the canyon.

Canyon system

The Valles Marineris begins in the west with a feature known as the Noctis Labyrinthus, an area of intersecting rift valleys that form a maze-like pattern. The opposite, eastern end is bounded by a chaotic terrain of irregular features, where smaller canyons and depressions give way to outflow canyons. These carried ancient rivers of water out of

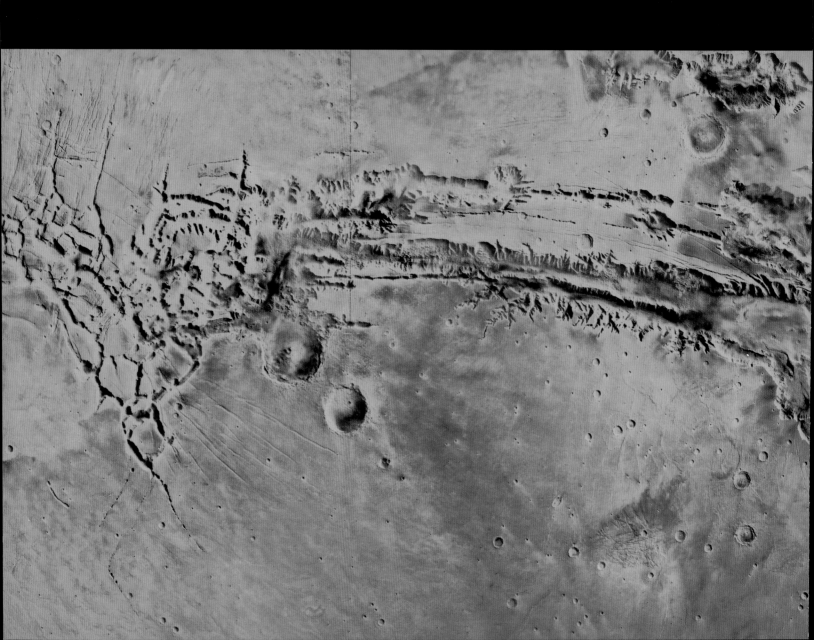

Marineris toward the lowland region to the north. This whole area has seen extensive water erosion; millions of cubic miles of material have been removed by water flow. The system's canyons are called chasma (plural: chasmata). The main chasma in the western part of the system is known as Ius. The central complex is made of three parallel canyons, Ophir, Candor, and Melas Chasma.

▶ **Melas Chasma**

This view of Melas Chasma is a still from a digital "fly-through" made by NASA from the available imagery of the Valles Marineris. Melas Chasma's depth suggests that it may be the best site for a manned outpost as it would have the highest natural air pressure on Mars.

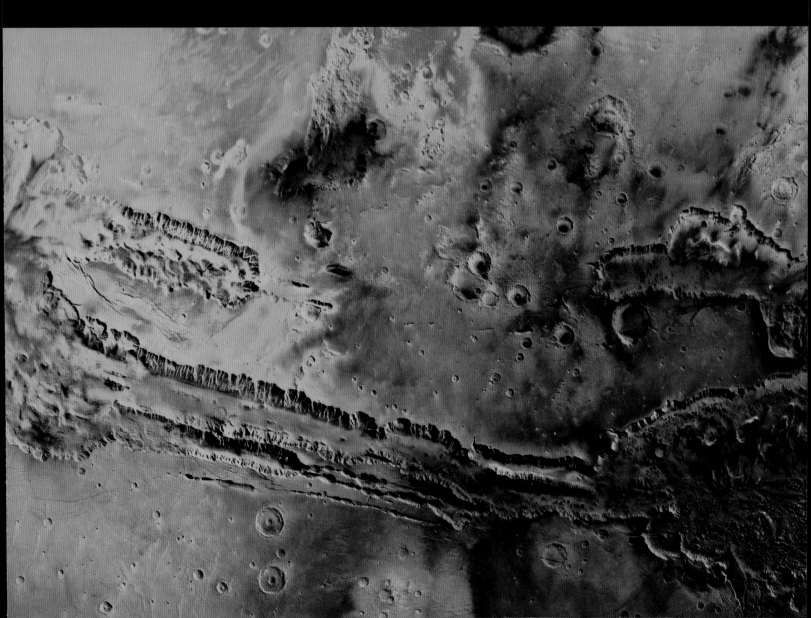

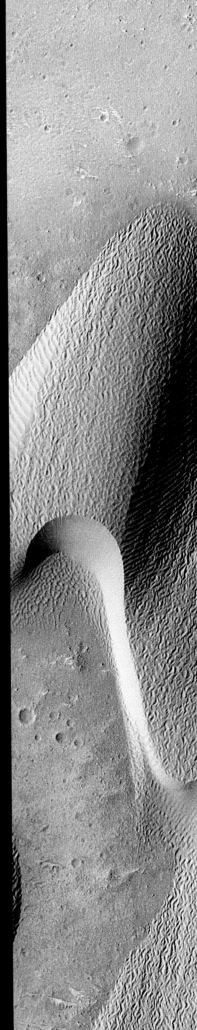

Seeing the surface from above

Spacecraft have been in orbit around Mars since 1971. In 2006, there were even four in orbit at the same time. Between them these Mars orbiters have photographed and mapped the entire globe. They have discovered surface ice, new craters, and tectonic features, evidence of water erosion, and unusual surface weather patterns.

The Mars orbiters

Mars was the first planet to be orbited by a spacecraft—NASA's Mariner 9. When the craft arrived at Mars in November 1971 it had to postpone its mapping program because the planet was enveloped in a giant dust storm, but it soon returned images that formed the first global map of Mars. From 1976, NASA's Viking 1 and 2 orbiters took thousands more images. Twenty years later, NASA's Mars Global Surveyor had completed nine successful years of orbits, sending back images of a much higher resolution.

Today, NASA's Mars Reconnaissance Orbiter (MRO) has been in orbit around the planet since March 2006. With its imaging spectrometer it is searching for minerals that formed through long-term interaction with water. The presence of such minerals would prove that liquid water once existed on Mars for a long period of time. The MRO is also investigating future landing sites by monitoring Mars's daily weather and surface conditions, and spot-imaging with its HiRISE camera (High Resolution Imaging Science Experiment), which can zoom in and photograph targets just 3 ft (1 m) across.

Dunes of Herschel Crater

his beautiful image taken by the Mars Reconnaissance Orbiter shows dunes on the floor of Herschel crater. The steep faces ("slipfaces") re oriented downwind, in the irection of motion of the dunes. and dunes form naturally as a esult of the movement of sand by he wind. The dunes in this image re somewhat crescent-shaped, but re being extended and distorted ownwind and merging with nearby unes; this complex behavior is ommon in dune fields on Earth.

Dust devil

dust devil casts a serpentine hadow over the Martian surface n this image acquired by the Mars econnaissance Orbiter's HiRISE amera. The devil is 2,600 ft (800) high and 100 ft (30 m) wide. It rose during a spring afternoon on he Amazonis Planitia region of orthern Mars in 2010.

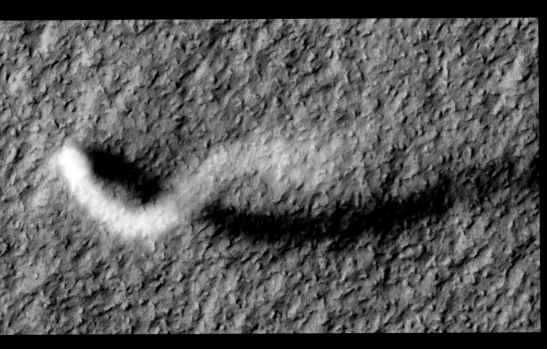

World of dunes

Sand dunes are extensive on Mars and have been photographed in many different regions. They were first seen when Mariner 9 investigated the planet from orbit in 1971–72. However, the resolution of the images sent back by the early orbiters was such that only two types of dune could be identified. These were barchan (arrowhead-shaped) and linear. The higher resolution images of the Mars Global Surveyor revealed several further types. Today, the superb quality of the images taken by the Mars Reconnaissance Orbiter reveals gullies, ripples, and many other features of the Martian dunes. Many look like the wind-blown sand dunes of Earth, which take on different shapes according to wind type and direction. Obstructions such as scarps and craters also influence the way sand collects and the shape it forms. Unlike Earth's dunes, which change continuously, Martian dunes appear to be static, and may have formed in the ancient past, either because the atmosphere was denser then or because dunes form very slowly in the thin atmosphere. Small fields of dunes are common in the center of impact craters, and have much smaller ridges of sand, called ripples. The dunes are composed of basaltic sand from eroded volcanic rock.

Seasonal liquid water?

With Mars's present low atmospheric pressure and low temperature, liquid water cannot exist on the surface except at the lowest elevations for a few hours. So a geological mystery arose in August 2011 when NASA's Mars Reconnaissance Orbiter revealed gully deposits that were not present ten years ago on craters in the southern hemisphere. Researchers hypothesized that these were caused by flowing salty water (brine) during the warmest months. The water flowed and then evaporated, possibly leaving a residue. However, the source of the water and the mechanism behind its motion are not understood.

Avalanches

Despite the generally static nature of Mars's landscape, dramatic rockfalls and avalanches do occur. A February 2008 HiRISE observation captured four huge avalanches in progress off a 2,300-ft (700-m) cliff near the Martian north pole. The reddish layers were rocks rich in water ice while the white layers were seasonal carbon dioxide frost. The landslide is thought to have originated from the uppermost red layer.

▲ Giant rockfall

In February 2008, tons of material, including fine-grained ice and dust and possibly large rocks, detached from a towering cliff near the Martian north pole and cascaded to the gentler slopes below. The cloud of dust was 590 ft (180 m) across and extended 620 ft (190 m) from the base of the cliff.

▶ Dry ice surfing

This 2013 image shows chunks of frozen carbon dioxide (dry ice) gliding down sand dunes on cushions of gas, similar to miniature hovercraft. They are plowing furrows known as linear gullies.

▶ Mystery water

A 2011 view of Newton Crater near Mars's equator shows warm-season features that may be evidence of liquid water flows. Scientists are not sure where the water came from or where it has gone.

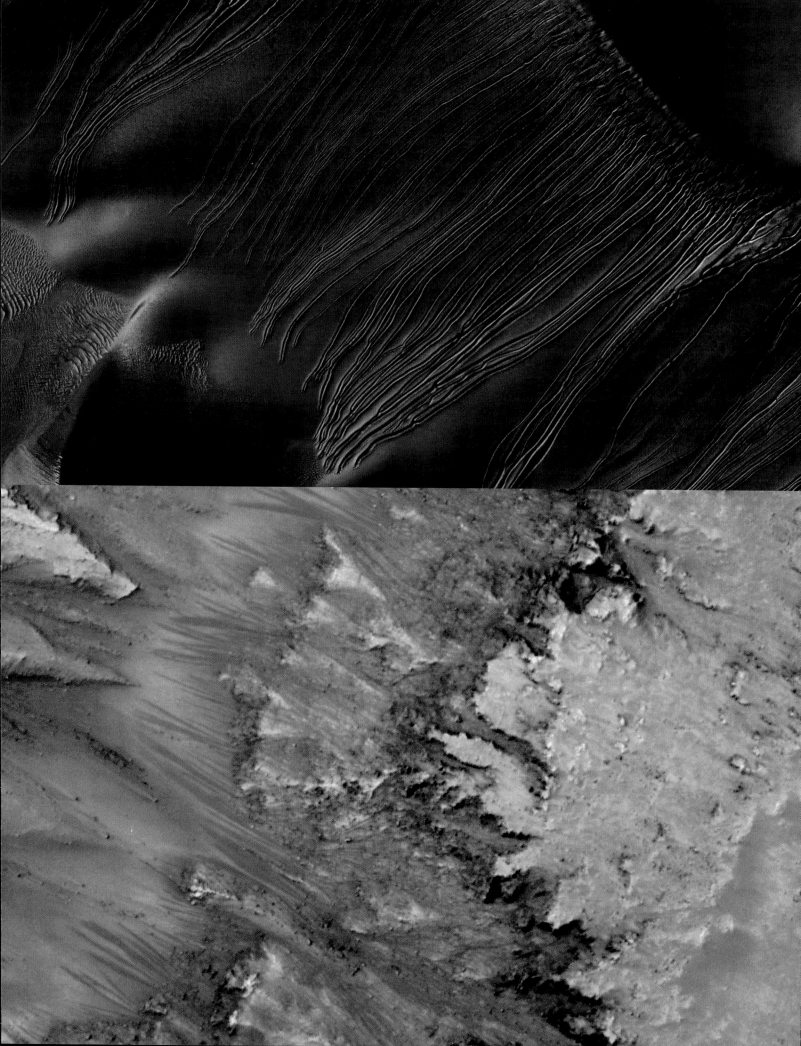

Mars landers

In all, six probes have made successful landings on the red planet. Three were designed to remain stationary where they landed; the others were designed to rove. The first rover was Sojourner which arrived on Mars with the Pathfinder lander in July 1997. It was small—about the size of a household breadmaker—and investigated sites near its landing area. The next two rovers, Spirit and Opportunity (which landed in 2004), were larger and much more sophisticated, carrying everything with them, including geology laboratories and communications equipment that did not rely on the landing craft. All these rovers were powered by solar panels.

Curiosity

The most recent and largest robotic rover, NASA's Curiosity, arrived in August 2012 and continues to trundle across the planet. Its goals include the investigation of the Martian climate and geology; assessment of whether the planet has ever had favorable environmental conditions for life, including an investigation of the role of water; and planetary habitability studies in preparation for future human exploration.

About the size of a small car, Curiosity is one of the most sophisticated scientific instruments ever built. It weighs 1,980 lb (899 kg), has a robotic arm equipped with cameras and a drill, and, unlike the previous rovers, is powered by a nuclear source, rather than by solar power. It carries a range of equipment, including a multi-spectra mast camera with zoom lenses; a "chemcam" for analyzing the composition of rock samples; two pairs of black-and-white navigation cameras mounted on the mast; an array of instruments to measure the environment, including humidity, pressure, temperature, wind speed, and ultraviolet radiation, as well as powerful communications equipment for sending its results back to Earth. Curiosity also has hazard avoidance cameras at its front and back so that it can autonomously avoid pitfalls and obstacles.

Life on Mars

For scientists, the great question Curiosity could answer concerns life on Mars. It seems unlikely that life exists there now, but did it exist there in the past? Will we find fossil evidence if we look hard enough and dig deep enough? These are questions to which we may soon have answers.

▲ Drilling holes

Curiosity's robotic drill has a dust-removal rotating wire brush that first cleans the rock area before drilling. The drilled rock dust is then scooped up and taken inside the rover for chemical analysis.

◄ Self portrait

Curiosity is here pictured in the Gale Crater on October 31, 2012, where its first rock sample collection took place. Its masthead camera is pointed at us. The photo was taken using its hand lens imager, one of 17 cameras mounted on the rover.

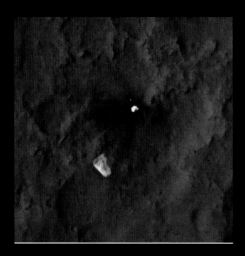

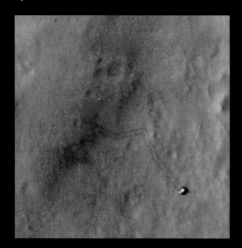

▼ Making tracks

Trundling over the Martian surface, Curiosity leaves tracks in the dust in this image taken by the Mars Reconnaissance Orbiter.

◄ Landing

This image taken by NASA's Mars Reconnaissance Orbiter shows Curiosity's landing site, in the Gale Crater, with the remains of Curiosity's protective shell in the center and its parachute to the lower left. Subsequent images show the parachute being moved around by winds in Mars's thin atmosphere.

▲ Lincoln penny

Part of Curiosity's camera calibration target is a 1909 Lincoln penny. Scientists use the calibration target to gauge the color and size of rocks and soil samples.

▼ 360° panorama

Curiosity used three cameras to take nearly 900 images over several days to produce this full-circle view of the Martian landscape. Looking south, this panorama was taken at the "Rocknest," a sand patch in the enormous Gale Crater.

▲ Sample taking

Held within Curiosity's scoop is a sample of powdered rock taken by the rover's drill. Once in the scoop, the sample is sieved and then delivered to the rover's rock and chemical analysis instruments.

▶ **Dry river bed**

The jagged sheets of rock in this image are the shattered remains of an ancient river bed. Evidence that this was once a river bed can be found in the rounded pebbles lying close to the bedrock remains.

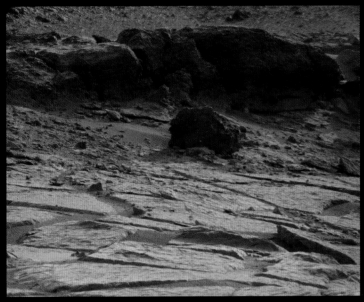

▲ **Point Lake outcrop**

One target for Curiosity in the Gale Crater was Point Lake outcrop. Measuring about 7 feet (2 m) wide and 20 inches (50 cm) high, its surface is covered in small cavities. A close-up image of these cavities (see top) shows that the holes are filled with material that is different from the outcrop's own rock.

JUPITER

The gas giant Jupiter dominates the Solar System. It is three times more massive than Saturn and has a total mass that is 2.5 times that of all the other planets put together. Even the Sun is only 1,047 times heavier. The planet is 5.2 times farther away from the Sun than Earth, and its surface is very cold. It takes just under twelve years to complete its orbit, and a little under 10 hours to complete each rotation. This rapid spin causes Jupiter to bulge, and it is 6.5 percent wider at the equator than at the poles. More than 60 moons have been discovered in orbit around the planet.

 Five spacecraft have investigated Jupiter. The most recent, Galileo, arrived in 1995. This orbiter circled the planet for the next eight years, sending back 14,000 images by the end of its mission. Galileo's instruments studied the atmosphere, magnetosphere, and moons. A probe was dropped into the surface, and transmitted data for 58 minutes. It reached a depth of about 120 miles (200 km) before it stopped transmitting.

▶ **Great Red Spot**

This view of Jupiter's Great Red Spot was taken by Voyager 1 in 1979. The image was taken through color filters and recombined to produce the color image. The giant storm is about twice the size of Earth. It is rotating counterclockwise, and has a high-pressure center. Winds in the outer regions of the storm are blowing at speeds of about 270 mph (430 km/h). Near the center, the eye of the storm is much calmer. Observations in 2010 suggest that the center may be significantly warmer and rotating in a clockwise direction.

▼ Turbulent atmosphere

The visible surface of the planet is Jupiter's colorful
atmosphere. The atmosphere is extremely turbulent,
and while the bands of weather systems and the
Great Red Spot persist over time, the details
within those bands are constantly changing.

Jupiter's atmosphere

The two main components of the atmosphere are hydrogen (86.4 percent of the molecules) and helium (13.6 percent). They are both transparent, just like the nitrogen and oxygen in Earth's atmosphere. The planet is shrouded by clouds composed of simple hydrogen compounds such as methane, ammonia, water, phosphine, and ethane. Solar heating at the equator makes the gases rise, allowing cooler polar gas to rush in. Convection currents carry gas upwards, creating distinct layers of cloud. The very top of the cloud layer, a hydrocarbon haze, has a temperature of about –256°F (–160°C). The temperature rises considerably toward the planet's core. Clouds of ammonia crystals, ammonium hydrosulfide, and water are visible on the descent through the top 75 miles (120 km). White zones of cool rising gas can be seen, and red-brown bands of warmer falling gas.

Stormy planet

Jupiter radiates about 1.7 times as much energy as it absorbs from the Sun. This excess energy is generated by gravitational contraction—the planet is very slowly shrinking. As a result of its strong internal heating, the temperature at the surface is almost uniform. The planet's rotational axis is tilted by just 3.1 degrees, which means that there are no significant seasons. However, the internal heat source is a major driver of the planet's weather. The rapid spin combines with the internal heat and solar heating to produce violent weather systems.

The planet spins once every 9.93 hours, faster than any other planet. This spin deflects north-south winds to the east or west to create cyclones and anticyclones, an effect known as the Coriolis effect, which also produces storms on Earth. The oval, cloud-like structures visible on the surface are giant storms. Many of these storms are short-lived, but some may last for decades, and the largest, the Great Red Spot, has been raging for at least 300 years. In 2005, a second large storm turned red. Unofficially dubbed Red Spot Junior, but officially know as Oval BA, it formed when three white storms merged with one another.

Jupiter's storms are often accompanied by lightning

▲ Thermal image

Detailed analysis of two continent-sized storms that erupted in Jupiter's atmosphere in March 2007 shows that Jupiter's internal heat plays a significant role in generating atmospheric disturbances. This infrared image of the storms—seen here as two bright plumes—was taken by NASA's Infrared Telescope Facility on April 5, 2007. The storm systems were triggered by water clouds moving rapidly up through the atmosphere. Scientists are studying the interplay between Jupiter's intense jet streams and other atmospheric phenomena such as storms. They hope to gain valuable insights into the way weather systems are driven on Earth, where jet streams dominate atmospheric circulation.

Atmospheric zones and belts

Jupiter's atmosphere can be divided into distinct zones and belts, each of which has its own unique characteristics. The wide pale Equatorial Zone is a relatively stable region. It is flanked by two dark equatorial belts known as the North Equatorial Belt and the South Equatorial Belt. These active regions periodically fade and reappear every 25 years or so. The South Tropical Region comprises the South Equatorial Belt and the South Tropical Zone. It is by far the most active region, and contains the Great Red Spot and many other weather systems. The alternating pattern of dark belts and pale zones is repeated as you move towards the poles, but above about 50 degrees latitude, they become less pronounced.

Great Red Spot

The most persistent feature on Jupiter through observations over the last 300 years has been the giant storm known as the Great Red Spot. It's latitude remains stable, varying over time by about one degree. However, its longitude is constantly changing. Defining longitude on Jupiter is not straightforward as its surface rotates at different speeds depending on latitude, but the Great Red Spot completes a lap around the planet about once every seven days.

The clouds in the Great Red Spot are colder than most of Jupiter's clouds, which means that they are at a higher altitude. Scientists calculate that it is about 5 miles (8 km) above the surrounding clouds. The reason for its reddish colour is not known. It may be caused by the presence of complex organic molecules, red phosphorus, or another sulfur compound. The dark red central region is a little warmer

▼ Bands of cloud

Detailed color maps of Jupiter were constructed from images taken by NASA's Cassini probe on December 11 and 12, 2000 as it passed by the planet on its way to Saturn. The smallest visible features in the maps are about 75 miles (120 km) across. The colors represent a very close approximation of the way the planet would appear to the human eye. The bluish-gray features along the northern edge of the bright central Equatorial Zone are hot spots. Small bright spots within the orange band north of the equator are thunderstorms.

than the rest of the outer areas. Periodically, the spot disappears entirely from the visible spectrum, but remains distinct at infrared wavelengths.

Hot spots

One of the least understood phenomena in Jupiter's atmosphere is the presence of hot spots. These are areas that are relatively free of cloud cover, where heat from the planet's interior can escape. The spots appear as bright spots in infrared images. The Galileo probe descended into a hot spot in the Equatorial Zone in an attempt to better understand them, but their origin is still unclear. They may be caused by local downdrafts of descending air, or they may be the result of waves that occur at the level of the whole planet.

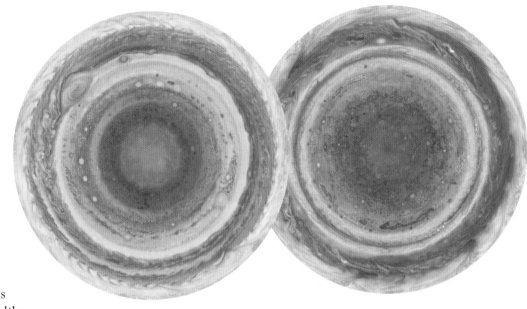

▲ Southern and northern hemispheres

These images of Jupiter's hemispheres were put together from data captured by Cassini. The southern hemisphere is on the left and the northern hemisphere on the right. The poles are in the center of the images while the edges are the equator. The polar regions are less clearly visible because Cassini was viewing them at an angle and peering through thick atmospheric haze.

SOLAR SYSTEM

▼ X-ray auroras

Jupiter's strong magnetic field produces spectacular
auroras at the poles that are bigger than the entire
Earth. This image shows X-ray auroras observed by
the Chandra X-ray Observatory, a space telescope
launched into orbit by NASA in 1999. These
have been overlaid on an optical image taken
from the Hubble Space Telescope.

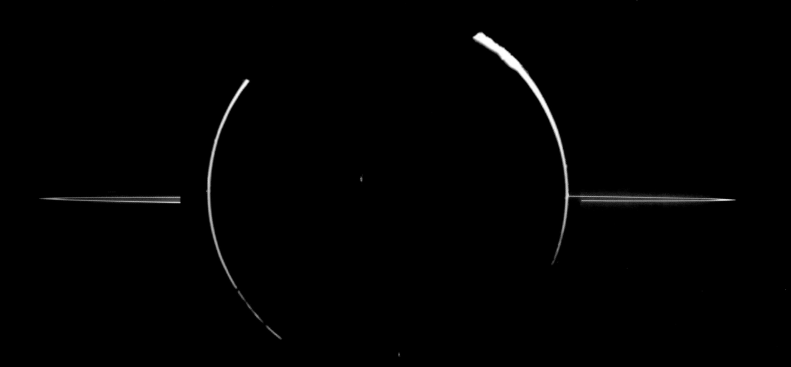

Jupiter's interior

Little is known about conditions in the interior of the planet. Our only clues are the average density and our physical models. It seems that the center of Jupiter has a temperature of about 43,000°F (24,000°C) and is at a pressure 50 million times greater than ground-level pressure on Earth. The central region consists of up to 15 Earth masses of rocky, metallic, icy material. On top of this is an extremely thick layer of liquid helium and metallic hydrogen. In the upper atmosphere, the hydrogen is molecular, each molecule consisting of two nuclei and two electrons (H_2).

Magnetic planet

In the pressurized hot interior, the electrons in the hydrogen have been stripped away and are shared between all the individual positively charged nuclei, making the layer highly conductive like a metal. The spinning, electricity-conducting liquid is stirred by convection currents, which produces mini-dynamos and a general planetary magnetic field. The magnetic field is the strongest of any planet at about 20,000 times the strength of that of Earth, and the magnetic axis is inclined at about 11 degrees to the spin axis. This field has a huge effect on ionized particles in the vicinity of the planet. There are radiation belts, and particles drop out of these to produce spectacular polar aurorae. The magnetosphere has a tail that extends about 370 million miles (600 million km)—to the orbit of Saturn.

▲ Jupiter's main ring

This mosaic of Jupiter's ring system was captured by Galileo when the Sun was behind the planet and the spacecraft was in Jupiter's shadow peering back toward the Sun. The small particles in the main ring and also the smallest particles in Jupiter's upper atmosphere stand out. The fainter gossamer ring and inner halo only become visible if the main ring is overexposed.

Rings

Jupiter's rings were discovered by Voyager 1 in 1979. The thin, faint system of rings is made of dust-sized particles that have been knocked off the planet's inner moons. The main ring is about 4,300 miles (7,000 km) wide but less than 18 miles (30 km) thick. Outside this is a flat ring about 520,000 miles (850,000 km) wide known as the gossamer ring, named after the two moons whose material makes up the dust, Amalthea and Thebe. On the inside edge of the main ring is a 12,000-mile (20,000-km) wide halo made of tiny dust particles that reaches down to the tops of Jupiter's clouds. The particles in Jupiter's rings are thought to have very brief lifespans numbering tens of years, and are constantly renewed by material eroded from the moons. The age of the system itself is not known, but it may date right back to the formation of the planet.

Comet impacts

Due to its strong gravitational pull, Jupiter receives more comet impacts than any other planet. Indeed, it may act as a shield to the inner planets, protecting them from comets. However, recent research suggests that Jupiter's gravitational pull affects the orbits of comets in the outer Kuiper Belt in such a way as to draw them toward the inner Solar System. It may also be that the presence of Jupiter does not significantly affect the number of comets that pass near to Earth and the other rocky planets.

Death of a comet

At some point in the 1960s, a comet, named Shoemaker-Levy 9, was caught by Jupiter's gravity and entered orbit around the planet. First detected in March 1993, Comet Shoemaker-Levy 9 had an unusual fragmented form, comprising a string of at least 21 chunks. The fragmentation had been caused in July 1992, when the comet passed close enough to the planet for tidal forces to rip it apart.

Between July 16 and July 24, 1994, the fragments of Comet Shoemaker-Levy 9 crashed into Jupiter's southern hemisphere one after the other. Each impact created a fireball of hot gas that was more easily visible from Earth than the Great Red Spot and lasted for several months. This was the first directly observed collision of objects in the Solar System and provided valuable data about the composition of Jupiter's atmosphere.

Recent impacts

Since 1994, several more fireballs have been observed. In 2009, an object between 600 and 1,500 feet (200 and 500 meters) in diameter crashed into the planet causing a fireball about the size of the Pacific Ocean. In 2010 and 2012, fireballs caused by the impacts of much smaller objects just a few meters across were detected. On each occasion, the impact was spotted by an amateur astronomer. Fireballs from these far smaller impacts only last for one or two seconds, so we may only be detecting a small proportion of the impacts that are taking place.

◀ **Shoemaker-Levy 9 heads for Jupiter**

This is a composite photo, assembled from separate images of Jupiter and comet Shoemaker-Levy 9 taken by the Hubble Space Telescope. The image of the comet, showing 21 fragments, was taken on May 17, 1994, two months before they collided with the planet. The image of Jupiter was taken a day later. The dark spot on Jupiter is the shadow of the moon Io.

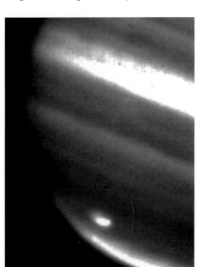

◀ **Impact scar**

This image shows a large impact on Jupiter's south polar region. It was captured on July 20, 2009, by NASA's Infrared Telescope Facility in Mauna Kea, Hawaii.

Jupiter's moons

At last count, Jupiter had 67 moons. The four
largest ones—the Galilean moons Io, Europa,
Ganymede, and Callisto—orbit close to the
planet and were formed in the equatorial plane
at the same time as the planet formed. Many
of the smaller moons were picked up later
on, when passing asteroids were caught by
Jupiter's gravitational field.

The Galilean planets are diverse worlds.
Io and Europa are closest to Jupiter and
are rocky bodies. Ganymede and Callisto
are mixtures of rock and ice. The orbits of
the inner three moons are in resonance
with one another. Io, the inner moon,
orbits Jupiter every 1.769 days. Next along,
Europa takes exactly twice as long as Io,
while Ganymede takes four times as long.
When Ganymede and Europa pass each other,
Io is always on the opposite side of the planet.

Io

Io is a little larger and denser than our Moon. It
orbits Jupiter at a height of 262,000 miles (421,600 km).
As it orbits, the moon is subjected to a strong gravitational

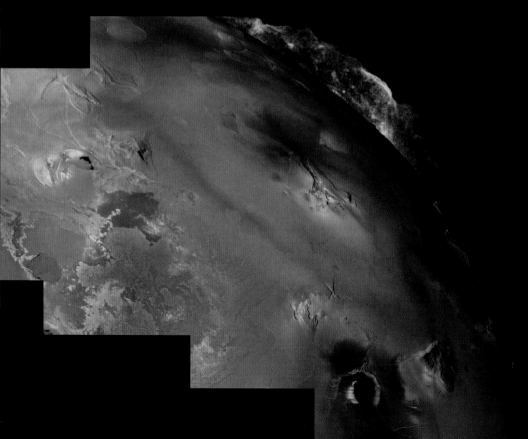

▲ Volcanic moon

In this color-enhanced image of Io taken by the
Galileo space probe, the dark spots on the surface
are active volcanos. At the center of the image is the
volcano Prometheus. A pale yellow ring encircles it.
The ring is made of sulfur-dioxide snow that was
deposited by the volcano's plume.

◀ Volcano erupting

A 180-mile (300-km) high plume rises from the
volcano Pele in this image of Io taken by Voyager 1
in 1979. Pele was the first active volcano to be found
on Io, and it was still active almost 20 years after this
picture was taken.

▶ **Conamara Chaos**

This view of the Conamara Chaos region on Europa indicates that the moon has been resurfaced relatively recently. The irregularly shaped blocks of ice were formed by the break up of the crust. The blocks were rotated, tipped, and partially submerged within warmer slush or water, before the surface froze solid again. This is a composite of images taken during Galileo's sixth orbit of Jupiter in February 1997, with insets of greater detail taken on December 16, 1997, when the probe was just 550 miles (880 km) above the surface.

pull from Jupiter on one side, and a weaker pull from Europa on the other. Changes in the relative strength and direction of Europa's pull cause Io's surface to flex. The flexing is accompanied by friction, which produces heat. This heat keeps part of the interior molten.

The molten interior erupts through the surface and constantly renews it. No impact craters can be seen on the moon's young surface. More than 80 active volcanoes have been identified on Io, and more than 300 vents. Super-heated sulfur dioxide shoots through fractures in the crust, creating plumes of cold gas. The material in the plumes falls back to the surface as snow and leaves oval-shaped frost deposits.

Europa

The smallest of the four Galilean moons is 0.65 times the mass of our Moon, and is one of the most intriguing objects in the Solar System. At the equator, the surface is about –225°F (–140°C), and towards the poles it is considerably colder. It shines brightly as the icy surface reflects five times as much light as our Moon. The surface is marked by impact craters, indicating that it is older than the surface of Io. The ice is

also marked by brown grooves and ridges, some of which are thousands of miles long. The mottled appearance is caused by parts of the crust that have broken up and floated to the surface. Dark spots called lenticulae formed when large globules of warmer slushy ice pushed up from under the surface ice and temporarily melted it.

Below the ice, which is just tens of miles thick, may lie a deep liquid sea. This layer of water is estimated to be 50–105 miles (80–170 km) thick and contain more water than all of Earth's oceans put together. Europa's oceans may contain all the necessary requirements for life to have developed there—liquid water, minerals, and a relatively stable climate. Below the liquid layer is a rocky mantle that surrounds a metallic core.

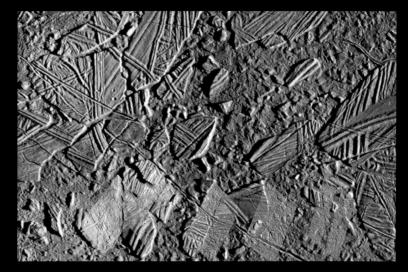

▲ **Europa**

This image of Europa was taken by Galileo in 1996. The bright feature with a dark central spot to the lower-right is a impact crater about 30 miles (50 km) across.

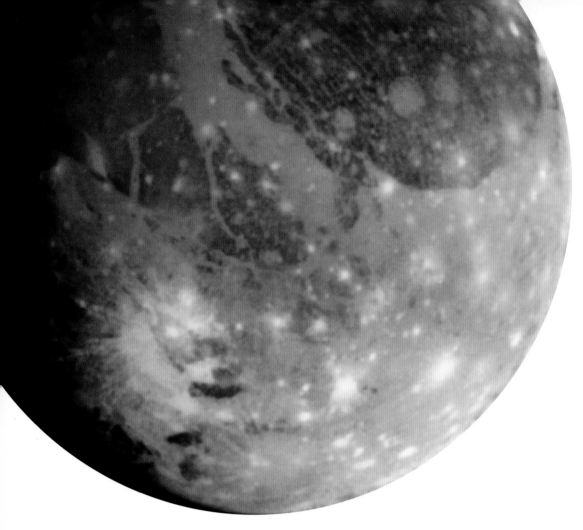

This natural-color view of Ganymede was taken by the Galileo spacecraft during its first encounter with the satellite. North is to the top of the picture, and the Sun is illuminating the surface from the right. The finest details that can be made out on this image are about 8 miles (13 km) across.

Ganymede

The third Galilean moon, Ganymede, is Jupiter's largest satellite. It is about twice the mass of the Moon and larger than both Pluto and Mercury. It is thought to be a mixture of about 60 percent rock and 40 percent ice. Ganymede's surface is a patchwork of dark and light areas. The dark regions are more highly cratered, indicating that they are older. The dark material is thought to be hydrated mineral clays. Long depressions, called furrows, about 4 miles (7 km) wide cross the dark areas. These may have formed when ice from under the surface flowed into new craters, dragging material across the surface. The bright terrain contains a high percentage of water ice with patches of carbon dioxide ice. It is smoother, but is crisscrossed by ridges and grooves. These were produced by tectonic stretching of the surface. The bright, circular areas are known as palimpsets. These are the smoothed-out and filled-in remains of craters formed in the distant past.

Ganymede is the only moon in the Solar System known to have a magnetosphere. This may be produced by currents in a layer of liquid water beneath the surface with dissolved electrolytes in it. The ocean may be buried at a depth of around 100 miles (160 km), sandwiched between layers of ice.

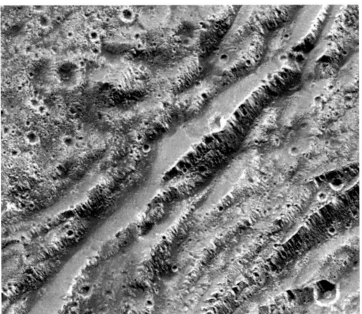

▲ **Faulted surface**

The Nicholson Regio on Ganymede is an ancient, heavily cratered dark area that is faulted by a series of scarps. The faulted blocks form a series of stairs like a tilted stack of books. Similar features are found on Earth when tectonic forces break the crust and blocks are pulled apart and rotated. Ganymede's faults may have been created by a similar process.

Callisto

The outermost, and darkest, of the Galilean moons, Callisto is still brighter than our Moon due to its highly reflective surface. Like Ganymede, it is thought to be about 60 percent rock and 40 percent ice. The heavily cratered surface shows little evidence of surface faulting or volcanic resurfacing, and Callisto has undergone little internal change since its formation. It is rockier toward the center and icier toward the crust.

Meteorite strikes have created broad basins around impact craters. One of the largest, the Valhalla Basin, is about 1,600 miles (2,600 km) across, and was probably formed by an impact early in the moon's history. The impact fractured the crust, allowing ice from below the surface to flood the area.

▲ Craters on Callisto

This image, taken by the Galileo probe on its ninth orbit of Jupiter, shows a heavily cratered region near Callisto's equator. The 30-mile (50-km) wide double ring crater in the center is called Har. The rounded mound is an unusual feature. It was probably created by the uplift of ice from below. Har is an older feature than the 12-mile (20-km) wide crater on its western rim. The image was taken when Galileo was 8,590 miles (14,080 km) from the moon.

◀ Ice and Rock

Taken in May 2001 by Galileo, this is the only complete global color image of Callisto. The surface is uniformly cratered all over, but is not uniform in color or brightness. The brighter areas are probably mainly ice, while the darker areas are highly eroded, ice-poor rock. The number of craters indicates that the surface of the moon has a long history without resurfacing.

SATURN

Saturn is 95 times more massive than Earth and 9.5 times farther away from the Sun. It is garlanded with a beautiful equatorial ring system, and orbited by a family of moons. The rings are inclined at 26.7 degrees to the orbit, which means that the top of the rings can be seen from Earth for half of Saturn's year—29.46 Earth years—and the underneath for the other half year. Saturn has a core similar to that of Jupiter, made of up to 15 Earth masses of rocky and icy material, and has slightly less hydrogen and helium in its atmosphere. Since 2004, the Cassini orbiter has been sending back information from its orbit around Saturn, revealing many of its secrets.

▶ **Beautiful planet**

This composite image of Saturn was put together from 102 pictures taken by the Cassini probe on October 6, 2004. As with Jupiter, all that is visible from Earth are the tops of the cloud layers.

▲ Saturn and Titan

This image was taken by Cassini from just to the north of the rings, and shows Saturn's largest moon, Titan, passing in front of the planet. As winter approaches in the southern hemisphere, Saturn starts to take on a bluish hue. This change in color is probably caused by a reduction in the intensity of ultraviolet light, which produces a haze during summer. As the haze clears, sunlight is scattered directly by the molecules in the atmosphere, and as on Earth, this makes the sky appear blue.

▶ **Heat image**

Three images taken in January 1998 by the Hubble telescope have been combined to produces this false-color image of Saturn. It shows the planet in reflected infrared light. The different colors indicate cloud layers at varying heights and chemical composition. The clouds are thought to consist mainly of ammonia ice crystals. The rings cast a shadow on the northern hemisphere. The bright stripe in the left side of the shadow just above the rings in this image is caused by sunlight streaming through a gap in the rings known as the Cassini Division.

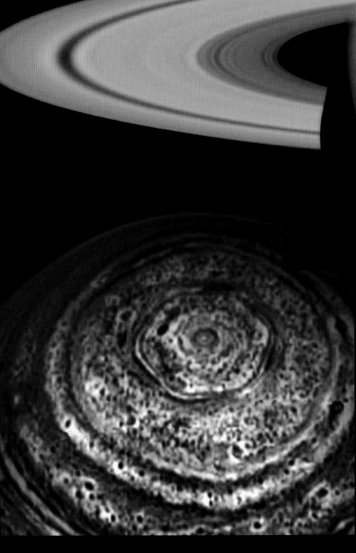

▲ **North pole vortex**

Taken by Cassini in 2006, this is the first image to capture the entire six-sided vortex at the north pole in one shot. It was taken during the dark northern winter and the image uses Saturn's own thermal glow as its light source—the planet glows at a wavelength of 5 microns, seven times the wavelength visible to the naked eye. The nested set of alternating white and dark hexagons that make up the vortex extends deep into the atmosphere.

Saturn's atmosphere

Saturn is colder and so less chemically active than Jupiter, which means that its clouds are less distinct and duller in color. Saturn was less effective at capturing gas from the preplanetary cloud than its inner neighbor, with the result that the Saturnian atmosphere appears comparatively calm. However, outbreaks of brighter clouds and spots are triggered by summer in the northern hemisphere. The white ammonia clouds are quickly spread around the planet, which has 1,100-mph (1,800-km/h) winds and a short spin period of 10.66 hours. The high-speed winds blow in the same direction as the planet's rotation. Bands of gas form in the upper atmosphere and these encircle the planet. Clouds and storms form within these bands. One region just below a latitude of 30 degrees south has been nicknamed "Storm Alley." In 2008, a storm raged there for seven months, generating bolts of lightning 10,000 times as powerful as lightning on Earth.

Polar regions

Saturn's polar regions are dominated by giant vortexes created as the planet's atmospheric bands spiral in towards the poles. The vortex at the north pole is hexagonal. It was first discovered by Voyager 1 in 1980, and was still there 26 years later when Cassini observed it. The vortex is a clearing in the clouds that extends more than 47 miles (75 km) upward. It is made up of a series of hexagonal bands centered on the pole, where it appears to be locked in place. At the south pole, the vortex is similar in appearance to a hurricane on Earth, with a clear "eye" about 900 miles (1,500 km) across surrounded by a wall of thick cloud. Winds blow around the ring at speeds of about 350 mph (550 km/h) Two arms of cloud spiral out from the central ring. As the seasons change, Cassini will monitor these weather systems to determine the role the planet's internal heat plays in powering them.

Magnetosphere

Like Jupiter, Saturn has a magnetic field and a magnetosphere, but these are smaller than its giant neighbor. The axis of the magnetic field is directly along the spin axis. This is puzzling because theories for field generation by dynamos inside a liquid ionized hydrogen core usually require there to be an angle between the two axes.

▶ **Strange aurora**

This view of an aurora over the north pole was produced by combining two images taken by Cassini. The aurora glows blue, while the clouds underneath it appear red. The aurora is unlike any previously seen in the Solar System. On Saturn, auroras can cover the whole of the pole, whereas auroras around Earth and Jupiter are usually confined to rings surrounding the magnetic poles. This phenomenon indicates that the charged particles streaming in from the Sun are being affected by previously unexpected magnetic forces.

Cassini-Huygens probe

Since 2004, we have made many new discoveries about Saturn, its rings, and its moons thanks to the Cassini-Huygens mission, which was launched in 1997 and entered planetary orbit on July 1, 2004.

Once the spacecraft was in orbit around Saturn, the Huygens probe was sent down to the surface of Saturn's largest moon, Titan, touching down in January 14, 2005. The lander took just under 2.5 hours to descend to the surface, sending back signals about the makeup of the atmosphere all the

has been extended to 2017, by which time it will have completed 290 orbits of the planet and more than 110 flybys past Titan. Cassini's instruments include telescopic cameras and spectrometers, which work in a range of wavelengths far beyond visible light to produce spectacular images revealing hitherto unsuspected features. Infrared images have revealed storms in Saturn's atmosphere and been used to make temperature maps of the rings. Ultraviolet imaging reveals the ice content of the rings and moons. Other instruments study the

▶ **Storm tail**

This false-color mosaic of images taken on January 12, 2011 shows a huge storm raging in Saturn's northern hemisphere, which Cassini monitored for several months. Red and orange colors indicate clouds deep in the atmosphere, yellow and green indicate mid-level clouds, and white and blue indicate high clouds and haze. The center of the storm is blue due to high haze overlying a hole in the deep clouds. Around the center, a ring of deep-level cloud forms into a tail.

▲ Saturn eclipse

Cassini looks back toward the Sun from behind Saturn. This view from the shadow reveals that the night side of Saturn is partly lit by light reflected from its rings. The rings themselves appear dark when silhouetted against the planet, but bright viewed away from it. During the eclipse, the rings were so bright that new ones were discovered. The pale blue dot to the left, just above the bright main rings, is Earth.

Saturn's rings

The rings of Saturn, one of the most beautiful sights in the Solar System, can be seen through a small telescope. They extend from about 1.2 to about 2.3 times the radius of Saturn. First observed 400 years ago, the rings were initially thought to be solid. In fact, they consist of billions of orbiting bodies ranging in size from grains of sand to bus-sized rocks. Every object in the rings is covered with ice, which reflects about 60 percent of the sunlight. Each body has an independent orbit. Those on the inner edge go around the planet every 5.6 hours, while those on the outer edge orbit every 14.2 hours. The thickness of the rings is about 65 feet (20 m), the size of the largest body.

Seven of the rings are identified by a letter, given to them in order of discovery, meaning that D is in fact closest to the planet. The three main rings, A, B, and C, are the most easily seen. Beyond these lie F, G, and the diffuse E ring. In 2009, the Spitzer Space Telescope discovered a faint donut-shaped dust ring far beyond the other rings. The dust ring starts about 4 million miles (6 million km) from Saturn and extends twice as far again.

Changing system

Saturn's rings are a dynamic, ever-changing system. The planet's moons interact with the particles, their gravity maintaining the gaps between the rings. At certain distances from the planet the density drops dramatically. Between the A and B ring lies the 2,200-mile (3,500-km) wide Cassini Division, a gap caused by the satellite Mimas.

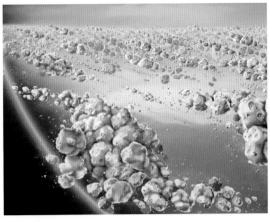

▲ Particle clumps

An artist's impression of the rings shows elongated clumps of particles with mostly empty space in between them. Clumps are constantly forming and breaking apart within the rings.

▼ Particle sizes

In 2005, Cassini sent three simultaneous radio signals through Saturn's rings to Earth. The changes in each signal as Cassini moved behind the rings were measured to calculate the distribution of the ring material. Purple color indicates regions where there are few particles smaller than 2 inches (5 cm) across. Green and blue indicate regions with more smaller particles. Boulder-sized particles are present in all the rings.

Any body at the inner edge of the gap orbits Saturn twice for every single orbit of Mimas. The gravitational field of Mimas effectively pushes or pulls particles out of the gap, forcing them to orbit Saturn in the outer or inner ring.

The moon Prometheus pulls at the F ring, tugging at its inner edge, while from the other side, Pandora pulls at its outer edge. The two small moons also affect each other's orbit, and the tug of Pandora's gravity periodically sends Prometheus crashing into the F ring. Orbiting in the 26-mile (42-km) wide Keeler Gap near the outside of the A ring, the moon Daphnis pulls particles at the edges of the gap into wave-like structures that extend at a perpendicular to the ring, reaching about half a mile (1 km) high.

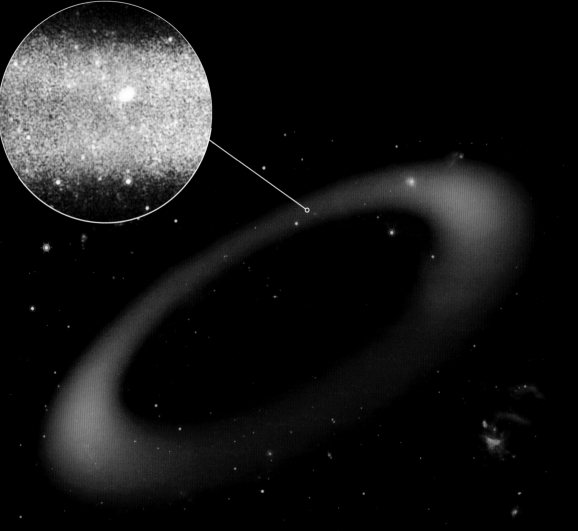

◄ Dust ring

This simulated infrared view shows the largest of Saturn's rings—the giant, nearly invisible dust ring discovered by NASA's Spitzer Space Telescope. Saturn appears as a small dot from outside the band of ice and dust. The bulk of the ring material starts about 4 million miles (6 million km) from the planet and extends outward another 8 million miles (12 million km). The ring's diameter is equivalent to roughly 300 Saturns lined up side to side.

▼ Narrow-angle view

In this mosaic of images from Cassini, the probe is at an angle of about 4 degrees to the rings, at a distance of about 1.1 million miles (1.8 million km). The image reveals the diverse colors of the complex system, with gaps, gravitational resonances, and wave patterns all visible.

Clumps

Particles within Saturn's rings interact with one another. Clumps of particles within the A and B rings are constantly colliding. The Cassini orbiter has recently discovered propeller-shaped features in the A ring, which measure about 3 miles (5 km) across with moonlets a few tens of meters across in their center. Depending on the Sun's angle to the rings, thin fingers can also appear, stretching out across the rings like the spokes of a wheel.

The origin of Saturn's rings

Several theories exist as to the origin of Saturn's rings. They may be material left over from the nebular material from which the planet was formed. Alternatively, they may be the remains of a moon that was struck by a comet or meteorite or was torn apart by Saturn's gravity.

A recent theory put forward by astrophysicist R. M. Canup suggests that the rings are the remains of the icy mantle of a large moon that was stripped of its outer layer as it spiraled toward Saturn. This would have taken place as Saturn was forming, when the planet was still surrounded by a gaseous nebula. The rings would have originally been about 1,000 times more massive than they are now. Over time, the outer portions would have coalesced to create Saturn's moons. This theory explains the predominance of ice and lack of rocky material in the rings and also in the composition of many of the moons.

Saturn's moons

Saturn has more than 60 moons, which vary greatly in size and distance from the planet. Some orbit within the rings, while others are far outside them. Most are small and irregularly shaped.

There are seven major moons, which are considerably larger than the others and are all round in shape. In order of size, they are: Titan, Rhea, Iapetus, Dione, Tethys, Enceladus, and Mimas. They all orbit Saturn in the equatorial plane. The inner five moons orbit within the rings, while Titan and Iapetus orbit outside the rings. The major moons have all been imaged by spacecraft, revealing cold worlds of rock and ice. Only Titan has an atmosphere, with a methane cycle that is reminiscent of Earth's water cycle.

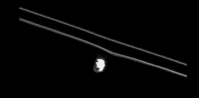

▲ Ring pull

The moon Prometheus gives the F ring a gravitational tug. The small moon is just 85 miles (136 km) wide, but it is close enough to the inner edge of the ring to pull the smaller particles toward it.

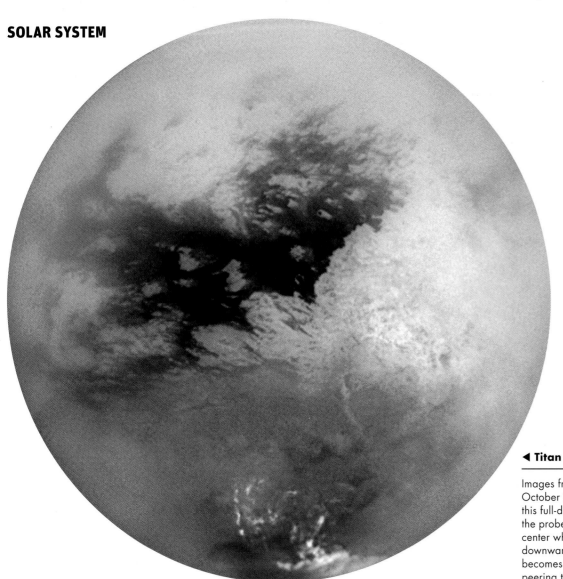

▲ Titan's atmosphere

Sunlight is scattered through Titan's atmosphere to form a ring of color as Cassini looks toward the night side of the moon. The north polar hood is at the top of this view, while the south polar vortex is just visible at the bottom.

◄ Titan

Images from Cassini's first flyby of Titan on October 26, 2004 were combined to produce this full-disc view. The Sun was directly behind the probe. The features are sharpest in the center where the probe was looking directly downward, while toward the edge, the image becomes increasingly fuzzy as the craft is peering through atmospheric haze.

Titan

Saturn's largest satellite, Titan, lives up to its name. With a diameter of 3,200 miles (5,150 km), it is bigger than Mercury and only a little smaller than the largest moon in the Solar System, Jupiter's Ganymede. Because of the speed at which its material came together, Saturn, like Jupiter, would have warmed up so much when it first formed that it glowed. This would have heated up the surrounding cloud, which explains why the ice in each of the condensing satellites tends to increase the farther away the satellite is from the planet. Titan is 760,000 miles (1.2 million km) away from Saturn, so its ice was gas rich. This has led to Titan being veiled by an extensive, smoggy, yellow-orange atmosphere made up of nitrogen, methane, and ammonia.

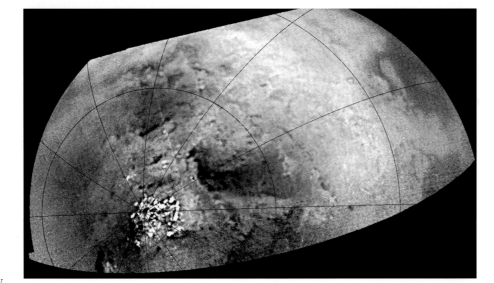

▲ Polar clouds

The bright features over Titan's south pole in this picture are clouds. It is a mosaic of images taken by Cassini on July 2, 2004, as the probe flew to within 211,000 miles (340,000 km) of the moon, its first distant encounter with it.

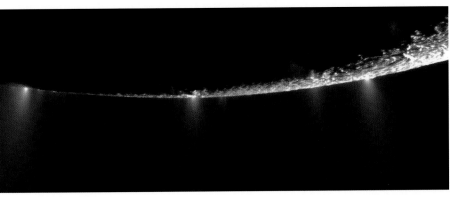

◄ **Ice plumes**

Dramatic plumes spray water ice out from many locations along the Tiger Stripes near the south pole of the moon Enceladus.

▼ **Tiger Stripes**

In this image, the Tiger Stripes on Enceladus are colored a false blue. The ice that spews out from these fissures, together with the presence of sodium in dust captured by Cassini, suggest that there may be underground oceans of salt water, which could contain the conditions for life.

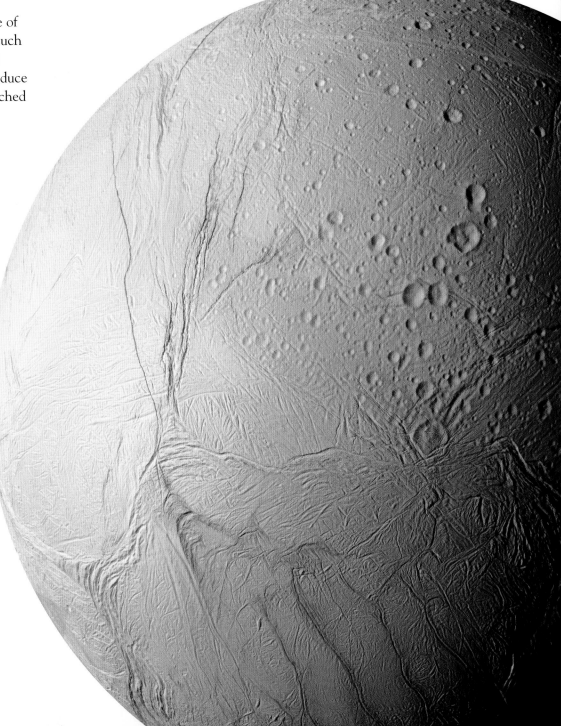

Methane rain on Titan

The surface of Titan has a temperature of –290°F (–180°C). The temperature is such that methane clouds form in the lower atmosphere, and these periodically produce methane rain. The Huygens probe touched down on the surface of Titan in 2005, revealing a gloomy world of hills and valleys interspersed with rivers and lakes of liquid methane.

Enceladus

Reflecting more than 90 percent of the sunlight that hits it, the icy moon Enceladus is one of the brightest objects in the Solar System. It measures 318 miles (512 km) across, and orbits within the E ring. Cassini has revealed that Enceladus is a volcanically active world and that large parts of it have recently been resurfaced. Near the south pole are four fissures known as the "Tiger Stripes," where ice particles, water vapor, and organic compounds erupt through the surface. The temperature in the fissures reaches –135°F (–93°C), compared to –330°F (–201°C) in the surrounding area. These eruptions give the moon a temporary atmosphere before the material leaves its orbit to join the E ring.

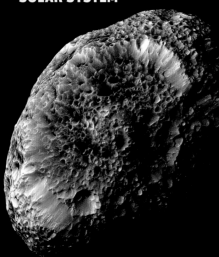

◀ Spongy Hyperion

Cassini swooped past the sponge-textured moon Hyperion in 2005 and again in 2010. This image from 2005 reveals a strange world strewn with deep craters. At the bottom of most of the craters is an unknown dark material. Hyperion is about 150 miles (250 km) across and rotates chaotically. Its density is so low that it may contain a large system of caverns in its interior. Due to its weak surface gravity, material at the surface has been blasted away rather than compressing back into the moon.

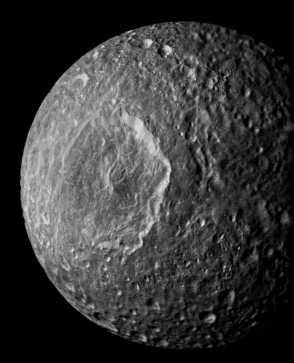

▲ Mimas

The innermost and smallest of Saturn's major moons, Mimas orbits within the rings and creates the Cassini Division between the A and B rings. Its average diameter is 247 miles (397 km), but it is around 19 miles (30 km) longer than it is wide. The icy surface is pitted with impact craters. The largest crater, called Herschel Crater, is 80 miles (130 km) wide and 6 miles (10 km) deep. Mountain peaks 4 miles (6 km) high are visible at the center of the crater.

Minor moons

Saturn's minor moons are divided into two groups: the inner moons and the outer moons. All but one of the inner moons orbit inside Titan. Only the sponge-shaped Hyperion, which has an eccentric orbit near to Titan, is farther from Saturn than its largest moon. The innermost of all, Pan, is just 16 miles (26 km) across. Prometheus and Pandora orbit either side of the F ring, while just beyond the F ring are Epimetheus and Janus, which share almost the same orbit, about 30 miles (50 km) apart.

The outer moons follow orbits that take them far beyond Titan. The innermost of the outer moons, Kiviug, is nine times farther from Saturn than Titan, while the outermost, Fornjot, is 20 times farther, at 15.6 million miles (25.1 million km) from the planet. The largest of the outer moons, Phoebe, is just 140 miles (230 km) across,

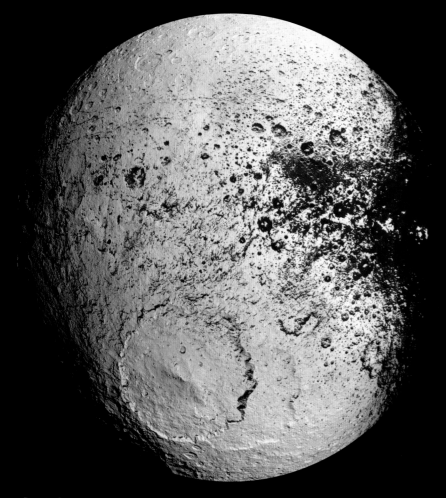

▲ Dusty Iapetus

Sections of the moon Iapetus are as dark as coal, while others are as bright as ice. This cold world of ice and rock is 892 miles (1,436 km) across and the most distant of Saturn's seven major moons, orbiting the planet every 79.3 days. The impact crater in the south is 313 miles (504 km) in diameter, and appears to be superimposed over an older crater of a similar size. The dark material may contain some form of carbon, and is less than 3 feet (1 m) thick. It is concentrated on the eastern part of the moon, and may be the remains of dust from other moons or from the dust ring that orbits in the opposite direction to Iapetus.

while many of the smaller moons are no more than 4 miles (6 km) across. These moons mostly display retrograde orbits, and are probably bodies from elsewhere in the Solar System that have been captured by Saturn's gravitational pull.

Discovering new moons

As of 2013, 62 moons have been discovered orbiting Saturn. Most of these have only been found since 2000, and the total number is likely to rise in the future as more small moons are discovered. The probe Cassini's wide-angle camera has been pointed at the rings to take images that can be searched for new moons. In this way, the 60th known moon, Anthe, was discovered between the orbits of Mimas and Enceladus in 2007. Tiny moonlets in orbit within the ring system that disturb nearby material in the rings may also be candidates for "moon" status in the future.

▲ Rhea passes in front of Saturn

Orbiting beyond the rings on the right of the image is the moon Rhea, which is 949 miles (1,527 km) across. On the left of the image, another moon, Tethys, casts its shadow on the planet.

◀ Ice moon Tethys

Measuring 666 miles (1,072 km) across, Tethys orbits in the ring system, taking 45 hours to circle the plane. This high-resolution image of an entire hemisphere of the moon was captured by Cassini in 2005. The white color of the surface is thought to be caused by fresh ice particles continually falling onto the moon from the E ring. A rift runs diagonally down the moon from the middle. Called Ithaca Chasma, this huge canyon may have been created when the interior of the moon froze, cracking the surface. This suggests that Tethys may once have contained subterranean oceans.

URANUS

Uranus is the third largest planet in Solar System and lies twice as far from the Sun as its neighbor, Saturn. It is pale blue and almost featureless, with a sparse ring system and a large family of moons. The planet is tipped on its side, which means that from Earth the moons and rings appear to encircle it from top to bottom.

The planet was the first to be discovered by telescope. It was spotted in March 1781 when the German-born British musician and amateur astronomer William Herschel was testing a new telescope at his home in Bath, England. Uranus is 19 times farther away from the Sun than Earth, so Herschel's discovery doubled the size of the known Solar System at a single stroke. It also made Herschel very famous; he was able to give up his job as a musician, was appointed Astronomer Royal to King George III, and became a professional astronomer.

◄ Hidden by haze

This image of Uranus, taken by Voyager 2 in 1986, was our first view of the planet close-up. It is so similar in size and composition to Neptune that astronomers sometimes put them in a separate category of planets known as "ice giants." Uranus's bland appearance indicates the presence of high-altitude methane clouds, which in turn overlay deeper clouds of hydrogen sulfide and ammonia.

▶ Planet revealed

This false-color, infrared image was taken by the Hubble Space Telescope in 1998. Uranus's ring system, vertical structure, cloud bands, and a number of its 27 orbiting moons can be seen clearly. The bright spots on the right side of the planet are springtime storms.

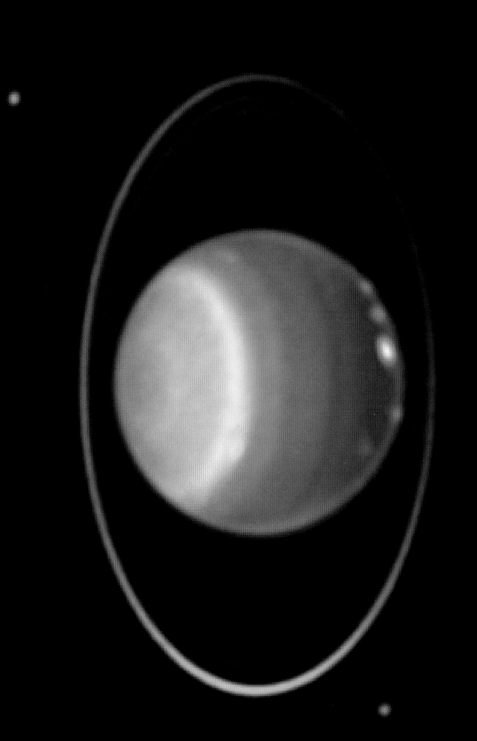

Composition

The U.S. Voyager 2 spacecraft flew by the planet in 1986, revealing a featureless globe. The atmosphere is 82.5 percent hydrogen, 15.2 percent helium, and 2.3 percent methane. It is the methane that gives the planet its pale color and haze. Interestingly, when the Hubble Space Telescope looked at the planet in 1997, a string of bright methane clouds had appeared, together with a patch of clouds over the south pole.

Uranus is smaller and less massive than Jupiter, but has the same average density. The interior material is less compressed, meaning that the planet consists of a higher proportion of rock, water, and ammonia ices, with a core of rock and possibly ice.

Spin and orbit

Uranus rotates every 17.24 hours. The spin axis is tilted at 98 degrees from the vertical to the orbital plane. This means that Uranus is tipped on its side and "rolls" around its orbit once every 84 years. It spin is retrograde, spinning in the opposite direction to most planets. This is probably the result of a collision with a planet-sized body when Uranus was young. Each of the poles points to the Sun for 21 years at a time, during periods centered on the solstices. In 1985, the Sun was above its south pole. It set there in 2007 and will not return to the polar sky for 42 years. When Voyager passed Uranus in 1986, its south pole was pointing directly at the Sun.

Planetary rings

The rings of Uranus were discovered on March 10, 1977. In the evening of that day, the planet passed in front of a star (SAO 158687). Observers were using the rate at which the starlight disappeared, and the duration of the eclipse, to investigate the varying density of Uranus's planetary atmosphere and to calculate a more exact value for the planetary diameter. To their surprise, the star was eclipsed not only by the planet but also by a series of 11 wispy, incomplete rings that extended out from 7,700 to 15,900 miles (12,400 to 25,600 km) above the planetary surface. The ring particles are much darker than those in the Saturnian ring system. The rings are not quite circular because Cordelia and Ophelia, two satellites 16 and 20 miles (26 and 32 km) across, are embedded in the ring system and perturb the ring particles.

▼ Dark rings

This close-up of Uranus's inner rings was taken by Voyager 2. All but the outer and inner rings are between 0.6 and 8 miles (1 km and 13 km) wide. They are so widely separated and narrow that the system has more gap than ring. The rings are made of charcoal-dark pieces of carbon-rich rocky material and dust particles.

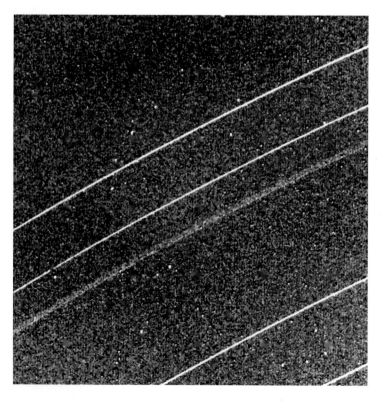

◄ Cloud patterns

In 1986, Voyager 2 photographed a southern hemisphere divided into two regions, with a bright polar cap (on the left) and dark equatorial bands. Clouds in Uranus's northern hemisphere are smaller, sharper, and brighter, and seem to be at a higher altitude. Recent observations have shown that the planet's cloud features have much in common with those of Neptune. In 2006, dark spots similar to those on Neptune were observed on Uranus for the first time (indicated by the white box).

► Mysterious spot

Scientists are still not entirely sure what has produced the dark spot in Uranus's northern (winter) hemisphere. One hypothesis suggests that it could be an anticyclonic vortex. The hemisphere has been in darkness for many years and the spot could be an indication of increased weather activity as the planet moves into its equinox.

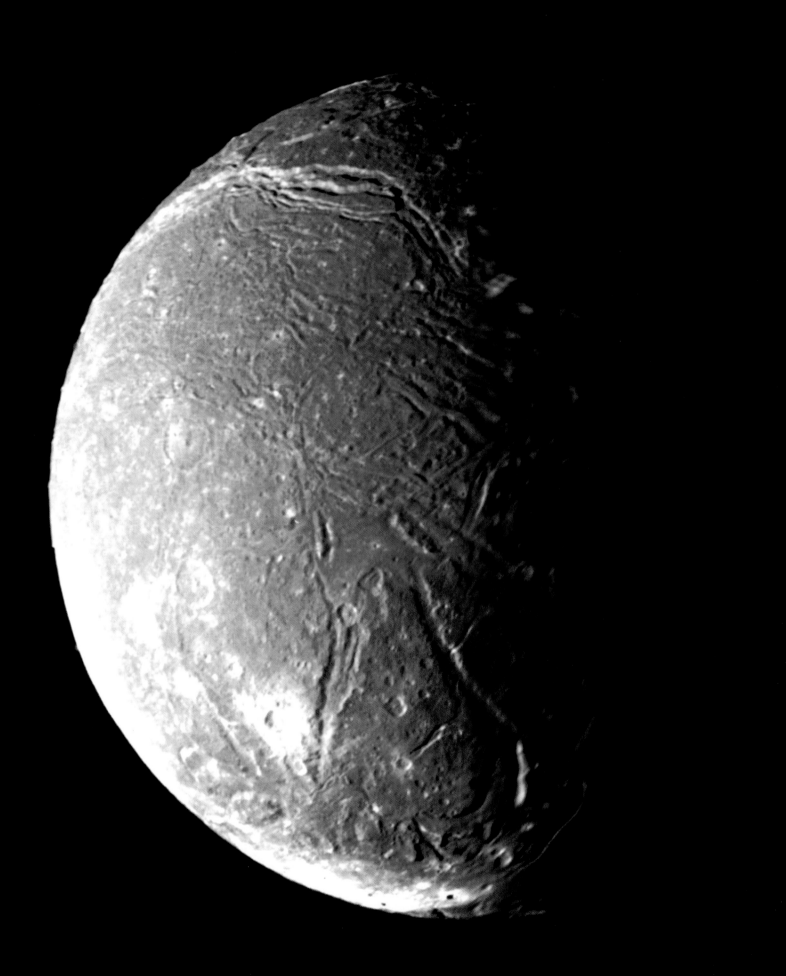

Uranus's moons

Uranus has 27 known moons. The two largest are Titania, which is 980 miles (1,580 km) wide, and Oberon, which is 946 miles (1,523 km) wide. They were discovered by William Herschel in 1787, a few years after he had discovered the planet. Umbriel, 727 miles (1,170 km) wide, and Ariel, 720 miles (1,158 km) wide, were found in 1851 by William Lassell, a British amateur astronomer. He was using a 24-in (60-cm) equatorially mounted reflector telescope at his observatory near Liverpool, England. A fifth moon, Miranda, was found in 1948 by Gerard Kuiper, the Dutch-American planetary expert. Ten more were identified when the Voyager 2 spacecraft passed Uranus. Since then, more moons have been found by detailed investigation of the Voyager images and by observations using the Hubble Space Telescope and powerful ground-based telescopes.

The densities of the largest satellites of Uranus indicate that they are all about half ice and half rock. Their surfaces reflect between 20 and 40 percent of the incident light. Craters abound and there are many fault-bound valleys and canyons.

Volcanic activity

Some of the regions on Uranus's moons have been resurfaced by icy volcanic material. Gravitational effects between the satellites has produced tidal distortions and tidal heating, triggering extended periods of volcanic activity. Miranda has fared the worst. Large polygonal darker areas on its surface have been produced relatively recently. Umbriel has also pulled Miranda into an orbit that is inclined at an angle of 4 degrees to the rest of the satellite system.

Only half of the surface of each satellite has been imaged. The complete satellite surface is illuminated by sunlight only when the planet is at its vernal or autumnal equinox. We are too late to take advantage of the 2007 equinox; the next occurs in 2049.

◀ Ariel

This composite of four images taken by Voyager 2 reveals that Ariel's surface is densely pitted with craters 3–6 miles (5–10 km) across. Valleys and fault scarps crisscross the pitted terrain. Extensive faulting appears to have occurred as the moon's crust has been stretched.

▲ Steep cliffs of Miranda

This narrow-angle image shows Miranda's varied surface. There are ridges and valleys, probably created by compression of tectonic plates. Cutting across them are numerous faults. The largest cliff, just below and to the right of the center of the image, is about 3 miles (5 km) high—higher than the walls of the Grand Canyon on Earth.

◀ Miranda's southern hemisphere

Nine images from Voyager 2 have been combined to produce this full-disk view of Miranda. The view is centered on the south pole and shows two different types of terrain. One is an old, heavily cratered terrain with relatively uniform reflectivity. The other is younger terrain with sets of bright and dark bands, scarps, and ridges.

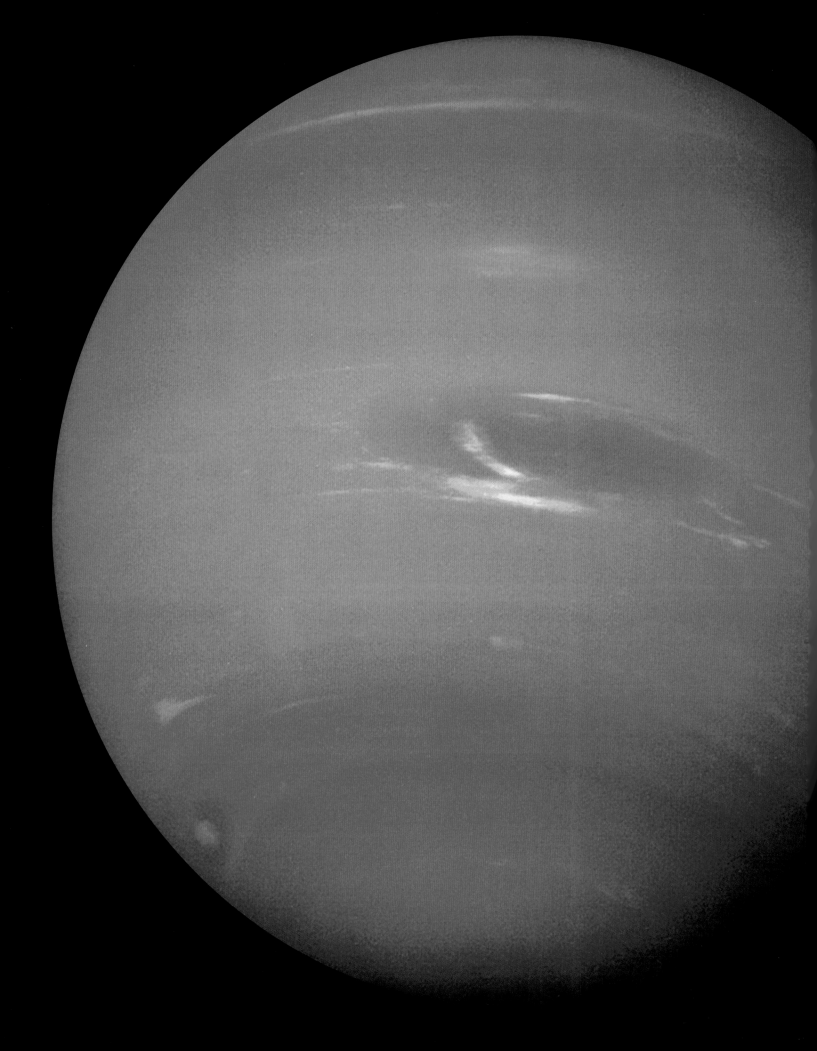

NEPTUNE

The outermost planet Neptune is 30 times farther from the Sun than Earth. It is one of the coldest places in the Solar System, with temperatures at the top of the clouds of –330°F (–201°C). Neptune was discovered by the German astronomer Johann Gottfried Galle on September 23, 1846. It had been noticed that the orbit of Uranus, itself only found in 1781, was being slightly pushed and pulled by another massive object. Using Newtonian gravitational theory, the position and brightness of the unknown object was predicted, and Galle discovered the new planet less than one degree from the position it had been calculated to be. Its discovery was a vindication for the universality of the phenomenon of gravitational force. Soon afterward, Neptune's largest moon, Triton, was discovered. So far, just one spacecraft, Voyager 2, has passed by the distant planet, sending back close-up images of the planet, its rings, and its moons. The probe passed by Neptune in 1989.

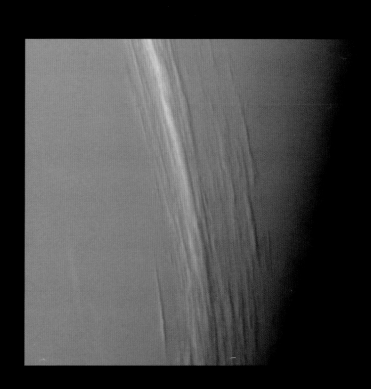

▶ Streaks of cloud

Taken by Voyager 2 two hours before its closest approach to Neptune, this image clearly shows the vertical relief of the bright clouds, which cast shadows on the side opposite to the Sun. The clouds are made of methane ice, stretching along lines of roughly equal latitude at an altitude of about 30 miles (50 km).

◀ Blue planet

Methane gas in Neptune's upper atmosphere absorbs the red wavelengths in sunlight, with the result that the planet appears a pale blue to the naked eye.

Cold and distant planet

Neptune is very similar in size and structure to Uranus. It takes 164.8 Earth years to orbit the Sun. From Neptune, the Sun appears 900 times fainter than it does from Earth. It spins every 16.11 hours, and has an equatorial plane that is inclined at 28.3 degrees to the orbital plane, which means that it experiences pronounced seasons. Only about 15 percent of its 17.1 Earth masses are in the form of hydrogen. Neptune has an extensive rocky and icy core, similar in mass to the cores of Jupiter, Saturn, and Uranus.

Neptune's atmosphere

During the formation process, the planet picked up less gas than its inner neighbors. The atmosphere is 79 percent hydrogen, 18 percent helium, and 3 percent methane. The percentage of methane is much higher than in the atmospheres of Jupiter and Saturn. Despite its outwardly benign appearance, the winds on Neptune are some of the fastest in the Solar System. An unknown internal heat source means that the planet radiates twice as much heat as it receives from the Sun, and this drives its weather.

Like Uranus, Neptune has a strange magnetic field as the magnetic dipole is offset from the center of the planet. The magnetic axis is inclined by 47 degrees to the spin axis. The field may be generated in the lower mantle region, where hydrogen, carbon, oxygen, and nitrogen compounds have become ionized by the high pressure so that they conduct electricity.

▼ Great Dark Spot

As Voyager 2 flew past Neptune, a large, dark anticyclonic storm system was seen in the southern hemisphere. This system is the same size as Earth and is reminiscent of the Great Red Spot on Jupiter. The dark spot is at a lower altitude than the surrounding blue methane. Also visible were a few fast-moving methane ice-crystal cirrus cloud features. Later imaging from Earth showed that the clouds remained but that the dark spot had disappeared.

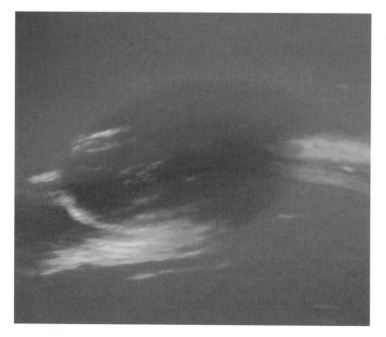

▶ Faint rings

Voyager 2 took this image of Neptune's rings in August 1989. In the outermost ring, 39,000 miles (63,000 km) out, material clumps into three bright arcs. The precise composition of the rings is not known, while the clumps are kept in place by the moon Galatea, which acts to shepherd the clumpy matter.

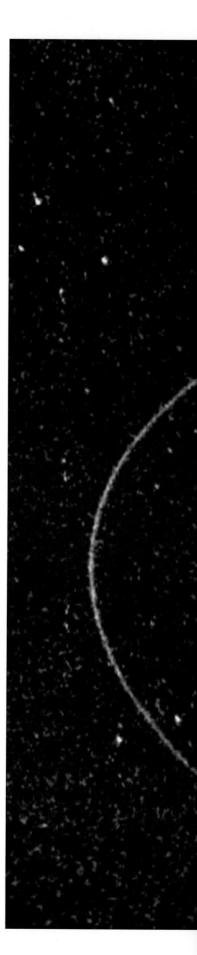

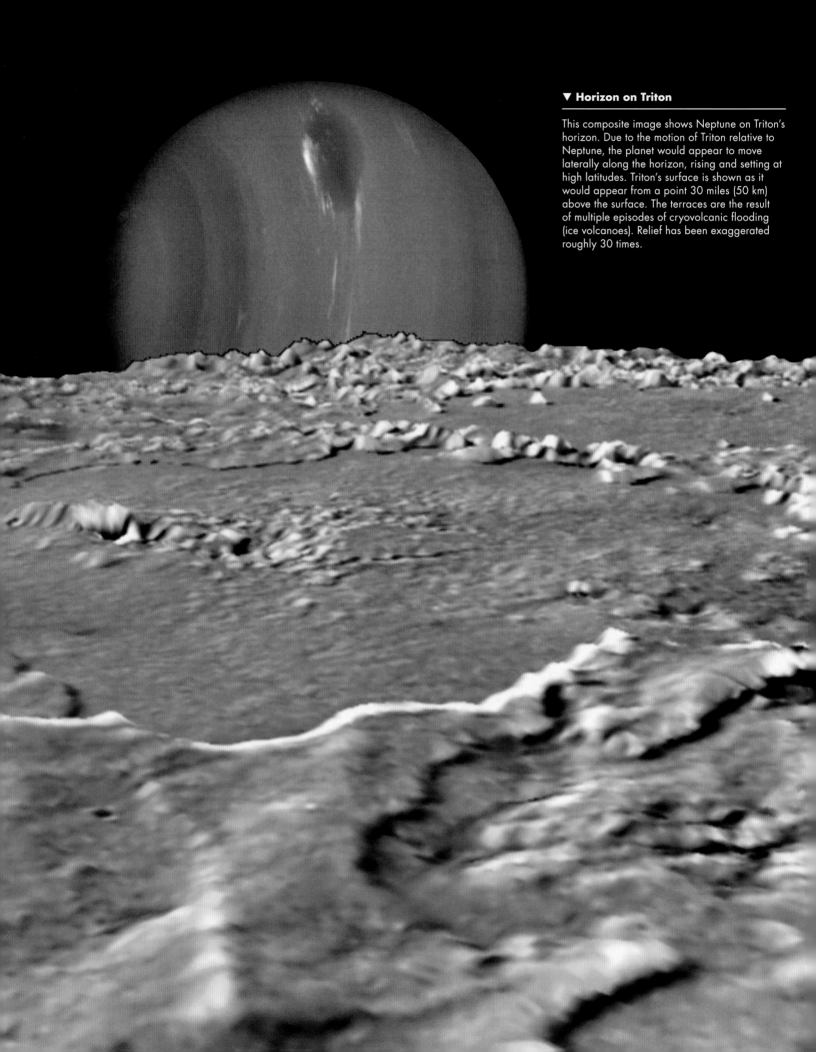

▼ Horizon on Triton

This composite image shows Neptune on Triton's horizon. Due to the motion of Triton relative to Neptune, the planet would appear to move laterally along the horizon, rising and setting at high latitudes. Triton's surface is shown as it would appear from a point 30 miles (50 km) above the surface. The terraces are the result of multiple episodes of cryovolcanic flooding (ice volcanoes). Relief has been exaggerated roughly 30 times.

Neptune's rings

Astronomers suspected that Neptune had rings several years before their existence was confirmed by Voyager 2 in 1989. Observations in the 1980s had shown stars' light blinking on and off just before they were obscured by the planet, providing indirect evidence of a ring system. As predicted, Voyager 2 found a series of thin rings: five main rings plus a sixth rather indistinct one. The rings are made of dust and small particles of unknown composition.

Family of moons

Neptune's largest moon, Triton, was found soon after the discovery of the planet. Another, Nereid, was discovered a century later orbiting beyond Triton. Data from Voyager 2's 1989 flyby led to the discovery of six more small moons orbiting closer to Neptune. Since 2002, five more moons have been found from Earth-based observations, all of them in distant orbits. The most distant, Neso, is 30 million miles (48 million km) from Neptune. Neso is the most distant moon in the Solar System, and takes 25.7 years to complete one orbit of Neptune.

Triton

By far the largest of Neptune's satellites, Triton is 1,680 miles (2,700 km) across. Like other large, close-orbiting moons in the Solar System, Triton is synchronously locked—it spins once during each of its orbits, and the same face always points toward the planet. Strangely, it is orbiting in the opposite direction to the spin of Neptune. Some think it may have been captured from the outer Solar System long after its adopted parent planet was formed.

Triton's surface

The temperature at the surface is about –400°F (–240°C), which means that the water, carbon dioxide, carbon monoxide, methane, and nitrogen are all frozen. Seasonal winds and winter nitrogen snow

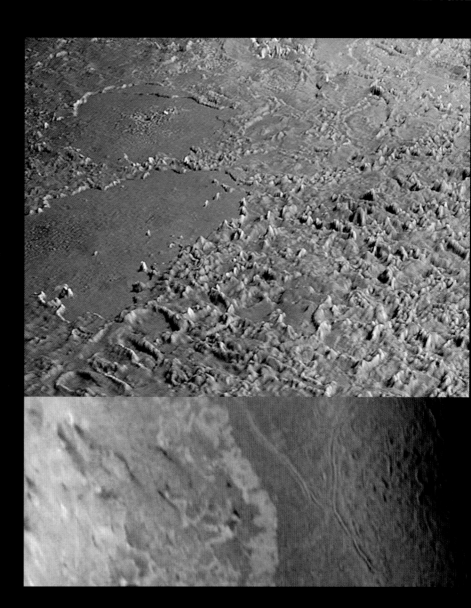

might help to account for the brightness of the surface. Away from the poles, Triton's surface is mottled with linear grooves and ridges interspersed with circular depressions. Resembling a melon skin, this surface has been named "cantaloupe terrain." The south pole is pinkish in color and marked with dark, dusty patches some 6 miles (10 km) wide and 90 miles (150 km) long. These are thought to have been blown out of nitrogen liquid and gas volcanic geysers, some of which remain active. The atmosphere of Triton is much more dense in the summer than in the winter. Its pink color is produced by sunlight reflecting off dirty, fresh methane ice.

▲ Surface features

These two views of the surface of Triton show its varied terrain. In the top image, the smooth volcanic plains were formed by icy lava. Parts of the surface have been eroded to form mounds and depressions. The bottom image is a false-color terrain map. The gray-blue area is an example of cantaloupe terrain. The parallel streaks may be particulate material from volcanic eruptions that has been blown in the same direction by winds in Triton's thin atmosphere. Triton's rugged surface, scarred by rising blobs of ice, faults, and volcanic pits, is very young and sparsely cratered.

▲ Voyager 2

The Voyager 2 space probe is currently in the outer reaches of the Solar System. It was designed with interstellar travel in mind, but its signal becomes fainter by a square of its distance from Earth and will eventually become too faint to detect. It carries a digital tape recorder to store data whenever it loses contact, which it then transmits when contact is reestablished.

Searching for new moons

A team of astronomers at observatories in Hawaii and Chile continue to search for new moons around Neptune. They do this by taking multiple images of the sky around the planet from both sites. The images are combined to boost the signal from any faint objects. In 2003, the team announced the discovery of three new moons, which showed up on their images as points of light against a background of stars that appeared as streaks of light. The search for small moons in distant orbit continues, and Neptune's family of 13 moons is likely to be added to in years to come.

Voyager 2

The Voyager 2 spacecraft gave us our first detailed look at Neptune and its main moon Triton, as well as confirming the existence of rings and discovering six additional moons. To date, it is the only spacecraft to have visited Uranus or Neptune.

The probe made its closest approach to Neptune on August 25, 1989. Its flight controllers then decided to take the opportunity to fly past Triton, regardless of the angle at which it would then fly away from Neptune. This meant that Triton was Voyager 2's final planned destination. It took a little less than 24 hours to move from Neptune to Triton, passing over the moon's north pole. On leaving Triton, the probe's trajectory was bent about 30 degrees south of the Solar System's plane of the elliptic. It continues to explore space, measuring magnetic fields and charged particles, and it is hoped that Voyager 2 will continue to send back radio messages until at least 2025, more than 48 years after it was launched. In about 296,000 years, Voyager 2 may pass 4.3 light years from the star Sirius.

On the chance that they are intercepted by intelligent life on their travels, both Voyager probes carry messages from Earth recorded on a phonograph record, known as the Golden Record. The record's contents were chosen by a committee headed by astronomer Carl Sagan.

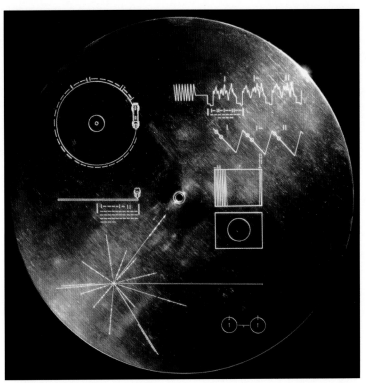

▶ Golden record

Etched into the Golden Record are instructions on how to play it written in binary code. The record contains images showing mathematical formulae, the Solar System, DNA, and human anatomy, plus images of animals, plants, and landscapes. It also contains recordings of human voices, greetings in 56 languages, and the songs of birds and whales.

▶ Comet Hale–Bopp

Comet Hale–Bopp was visible from Earth throughout 1997, passing closest to the Sun on April 1 of that year. It faded from view over the course of the rest of the year. The comet will next be visible from Earth in about 4385. It may be that a reference to a "long-haired star" on a pyramid in Saqqara, Egypt, is a record of this comet. The pyramid was made during the reign of Pharaoh Pepi I (2332–2283 BCE).

▼ Pluto

This artist's conception shows a view of the dwarf planet Pluto from the surface of one of its smaller moons. Pluto is the large disc in the center, while Charon, Pluto's largest moon, is to the right. The bright dot to the left is a second small moon in the system. Charon was discovered in 1978, while the smaller moons were first observed in 2005.

OTHER BODIES

There are trillions of minor bodies in the Solar System. Asteroids are irregularly shaped rocky or rocky metallic objects found mostly in a belt between the orbits of Mars and Jupiter. They range in size from dust and small boulders to city-sized objects. Beyond Neptune lies the Kuiper Belt, where objects are a mixture of rock and ice. Several dwarf planets orbit the Sun in the Kuiper Belt. Beyond that is the spherical Oort Cloud, in which trillions of icy comets orbit. The outer edge of the Oort Cloud may be about 1.6 to 3 light years away—nearly half way to the nearest star. Occasionally, comets are pulled out of their distant orbits toward the inner Solar System, where the Sun partially melts them to create their spectacular glowing tails.

This is an artist's conception of the largest dwarf planet, Eris, and its moon Dysnomia. The glowing object in the top left is the Sun as it would appear from this distance. Eris is about 96 times farther from the Sun than Earth and three times farther out than Pluto. It takes 557 years to orbit the Sun and, like Pluto, it has a highly eccentric orbit. Eris was only discovered in 2005. Compared to the reddish Pluto, which has bright patches, Eris is uniformly gray. It is far enough from the Sun that methane can condense evenly onto its surface, making it appear the same color all over.

Dwarf planets

A dwarf planet is a body that is large enough for gravity to have made it spherical, but not big enough to have cleared its own orbit around the Sun. It is failing this second test for a true planet that saw the downgrading of the second-largest object in the Kuiper Belt, Pluto, from a planet to a dwarf planet in 2006. Pluto has a highly eccentric orbit, and from 1979–1999 it was closer to the Sun than Neptune. It is locked into a 2:3 resonance with Neptune, meaning that twice the 248-year orbit of Pluto is equal to three times the 163.7-year orbit of Neptune. Pluto has several moons, including Charon, which has a mass of 15 percent that of Pluto, which itself has one quarter of the mass of our Moon. It is believed that Pluto and Charon may have formed at the same time when a single body broke into two.

So far, four other dwarf planets have been identified in the Solar System. The largest, Eris, was discovered in 2005. Ceres is the only dwarf planet in the Asteroid Belt, and the only one other than Pluto to have been observed well enough to confirm its status.

Comets

Before the advent of astronomical telescopes, about 10 comets were recorded per century. These were obvious to the naked eye, rather like two recent comets in Earth's skies, Hale-Bopp and Hyakutake. People were suspicious of comets. They appeared for a month or so, slowly changing their celestial position, and no one knew where they came from. Then, in 1687, English mathematician Isaac Newton illustrated how the orbital parameters of the Great Comet of 1680 could be calculated. A decade or so later, Newton's friend Edmond Halley showed that the comets of 1531, 1607, and 1682 were the same body. Today we know of about 160 short-period comets (orbital period less than 20 years) and 900 long-period ones (orbital periods greater than 20 years).

▲ Halley's Comet

Halley's Comet is visible from Earth every 76 years, and was seen over England in 1066. The Bayeau Tapestry tells the story of William of Normandy's defeat of Harold II of England at the Battle of Hastings that year. The comet appears on the tapestry with fearful Englishmen looking on. Contemporary accounts describe the comet as four times as bright as Venus. It passed to within about 1 million miles (1.5 million km) of Earth.

Comet nuclei

At the heart of a comet is an irregularly shaped dark nucleus, a cosmic dirty snowball a few miles across. The mix of dirt and snow has remained unchanged since the birth of the Solar System. Spacecraft have been sent to comets to study them, providing valuable clues to the Solar System's early history. In March 1986, the ESA's Giotto spacecraft passed to within 400 miles (600 km) of the nucleus of comet Halley. It was found to be 9.5 miles (15.3 km) long and 4.3 miles (7 km) wide, shaped like a potato. This was the first time anyone had seen the nucleus of a comet.

Inside a comet's nucleus is a deep-freeze mixture of "snow" and "dirt," the mass ratio between these being about 2:1. About 95 percent of the molecules in the snow are water, while the remaining 5 percent contains carbon dioxide, ammonia, and methane. We have very little idea what a comet is like inside, but unlike planet interiors, the temperatures have not been increased by radioactive decay, so there is no differentiation, and no variation is expected with depth. It is possible that the interior is full of holes, maybe because the smaller snowballs that came together to form the nucleus are very loosely packed. We do know that the nucleus is very weak, as was demonstrated by the ease with which the nucleus of comet Shoemaker-Levy 9 was pulled apart by tidal forces when it passed close to Jupiter in July 1992.

Eliptical orbit

Unlike the near-circular orbits of the planets, cometary paths are highly elliptical. For example, comet Halley comes close to the Sun every 76 years, passing between the orbits of Mercury and Venus. At the opposite end of its orbit, it is beyond Neptune, 35 times farther away from the Sun than Earth. When orbiting beyond Jupiter, Halley's nucleus is very cold, but near the Sun the snow just below the fragile, dusty surface is heated above –58°F (–50°C). As it warms, the ice begins to change into gas. The escaping gas exerts a pressure on the loosely held surface dust particles and they are pushed away. The nucleus is surrounded by a large, near-spherical cloud of escaping gas and dust, called a coma, which reflects sunlight. Comae can be as much as 60,000 miles (100,000 km) wide.

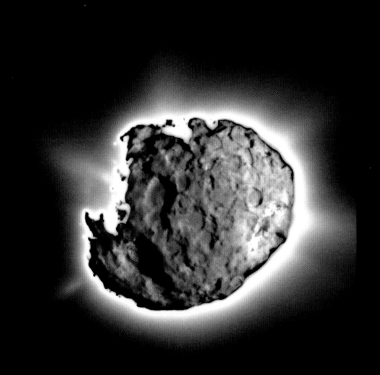

◄ Comet impact

This image of comet Tempel 1 was taken 67 seconds after Deep Impact's impactor spacecraft had been deliberately flown into it. The impact created a crater 300 ft (100 m) wide and 100 ft (30 m) deep. Scientists analyzed the resultant dust cloud and found that it contained more dust and less ice than they had expected, with a mix of clays, carbonates, silicate, and sodium. The comet was also found to be about 75% empty space. Future missions to comets will reveal whether or not this is a typical composition of a comet nucleus.

▶ Active surface

NASA's Stardust spacecraft flew by comet Wild 2 on January 2, 2004. Images taken by Stardust reveal an intensely active surface, with jets of gas and dust. Other than the Sun, Wild 2 has the most active known surface in the Solar System. Short-period comets such as Wild 2 have been captured in the inner

This image mosaic shows the southern hemisphere of the near-Earth asteroid Eros. The NEAR Shoemaker spacecraft touched down on its cratered terrain on February 12, 2001. Eros is an S-type asteroid 21 miles (33 km) long. Information from the NEAR mission shows that the asteroid is similar in composition to meteorites found on Earth, which are themselves pieces from similar S-type asteroids, possibly from Eros itself.

▶ **Crater on Eros**

The largest crater on Eros, called Psyche, is shown here from an altitude of 60 miles (100 km). It is 3.3 miles (5.3 km) wide. The crater provides clues to the history of Eros. Troughs and scarps cut through the crater, perhaps as a result of a large impact elsewhere on the asteroid. A large boulder is perched on the crater wall. Its position is the result of Eros's unusual gravity. Because of its uneven shape, the gravity "lows" on Eros are not necessarily at the lowest points of its craters.

Cometary tails

Pressure exerted by sunlight and the solar wind pushes gas and dust away from the coma in the opposite direction to the Sun, producing two huge tails. Near the Sun, cometary tails can be more than 60 million miles (100 million km) long. So, despite the small size of their nuclei, their comae and tails can make them very easy to see. The comets Tempel-Tuttle and Swift-Tuttle have the same orbits as the meteor showers, the November Leonids and the August Perseids. Large particles break away from the nucleus and slowly gain on or fall behind the comet, eventually forming a ring of meteoroid dust around the orbit. When Earth passes through this ring, a meteor shower occurs.

All the comets we see in our skies are dying. For instance, comet Halley typically loses a layer 8 feet (2.4 m) thick during each orbit. After 2,200 or so more orbits—170,000 years—Halley's comet will have disappeared.

Discovering asteroids

When the German astronomer Johannes Kepler tried to explain the orbital spacing of the planets using pure geometry, he realized that there was too large a gap between Mars and Jupiter. By the 1790s, astronomers were looking for a missing planet to explain the gap.

The first asteroid, Ceres, was discovered by Italian astronomer Giuseppe Piazzi in 1801 (Ceres is now called a dwarf planet). On March 28, 1802, Heinrich Olbers discovered another asteroid, which was named Pallas. Both Ceres and Pallas were surprisingly small, only a few hundred miles across. Their orbits nearly intersect. Olbers suggested that they were fragments of a much larger planet that had once orbited in the Mars–Jupiter region. He thought that this planet might have suffered an internal explosion or a cometary impact many millions of years before.

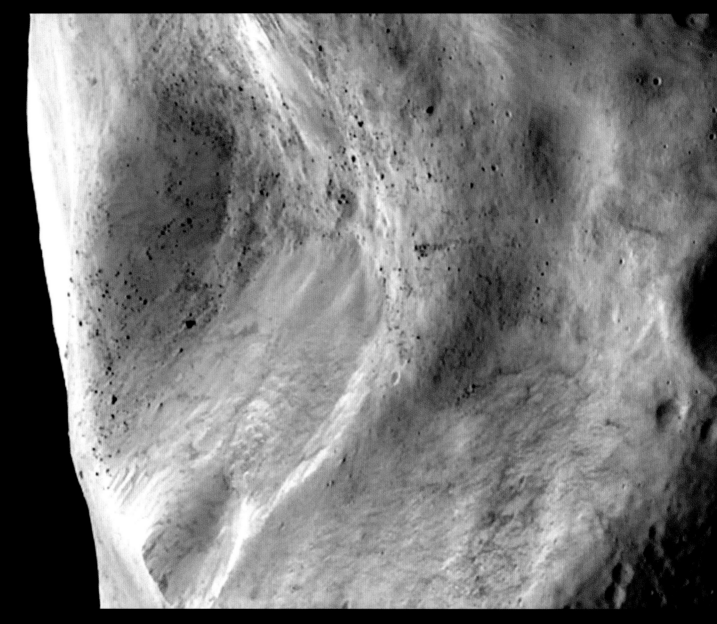

▲ Landslide on Lutetia

The ESA's Rosetta probe passed by the main-belt asteroid Lutetia in 2010. Lutetia is about 60 miles (100 km) in diameter and is heavily cratered. The largest impact crater is 30 miles (45 km) wide. The surface of the asteroid is covered in a layer of debris about 0.5 mile (1 km) thick. Landslides are thought to have been caused by vibrations created by impacts elsewhere on the asteroid. Past impacts have torn pieces out of the asteroid, which astronomers believe was once nearly spherical in shape.

Finding more asteroids

As the 19th century progressed, the introduction of larger telescopes and the use of photography enabled fainter, and thus smaller, asteroids to be discovered. The floodgates opened, and by 1900 about 500 asteroids were known. Today, large telescopes are used to hunt for asteroids, and tens of thousands now have well-known orbits and brightnesses.

The differing colors and reflectivities of asteroids suggested that some are rocky while others are metallic. They are far too small to

have atmospheres. Those larger than about 180 miles (300 km) wide are spherical, as their interior strength is insufficient to resist the pull of gravity. Smaller ones are irregular in shape, and the amount of light they reflect varies as they spin and present differing areas to the Sun. Through a telescope, they appear like moving stars. A typical asteroid in the Main Belt travels about a quarter of a degree per day against the distant starry background. Asteroids are discovered when the same area of the sky is observed night after night and the movement is detected.

Asteroid distribution

The Asteroid Belt has evolved as asteroids have smashed into each other, producing a host of fragments. The current size distribution is such that there are about 1,000 asteroids that are bigger than 58 miles (94 km), one million bigger than 5.8 miles (9.4 km), one billion bigger than 0.58 mile (0.94 km), and so on. Variations in the amount of light reflected from them indicate that most asteroids have spin periods of between 6 and 13 hours.

As more asteroidal orbits became known, it was realized that there were gaps in the distribution of their orbits. These gaps are caused by the gravitational influence of Jupiter and are equivalent to 1:3, 2:5, 3:7, 1:2, and 3:5 resonances with that planet. (In a 1:3 resonance, the asteroid orbits the Sun three times for every one orbit by Jupiter.) These resonances can lead to asteroids being thrown out of the belt. When this occurs, they tend to wander off through the Solar System, where they can hit planets and satellites. In 1918, Kiyotsugu Hirayama realized that some asteroids had very similar orbits. These asteroidal "families" had been produced by recent breakups of larger members of the main belt.

In the early 1970s, astronomers started to measure the brightnesses of various asteroids at different wavelengths. This led to a classification system based on surface color.

The most common three classes were labeled C, S, and M. Carbonaceous asteroids (C-type) represent about 75 percent of all asteroids and reflect only about 7 percent of the light that falls on them. As the name implies, their rock contains carbon. Stony, or silicaceous, asteroids (S-type) are three times more reflective. Rarer M-type asteroids are metallic.

Clues from meteorites

Meteorites are the remnants of small asteroids that hit Earth's atmosphere at velocities of about 12 miles (20 km) per second. Friction slows down the incoming rock and erodes away much of its surface. What is left falls to the ground. Meteorites provide a glimpse of what was once in the interior of planet-sized bodies.

At the dawn of the Solar System, there were about four Earth masses of rocky-metallic material in the Asteroid Belt. This was in the process of coming together to form a single terrestrial planet. However, Jupiter was on the other side of the snow line, meaning that it grew faster. Gravitational perturbation by Jupiter distorted the early asteroidal orbits, which were originally nearly circular, into elongated ellipses. This caused the original planetesimals to collide and fragment. There is now only about one part in 2,000 of the region's original mass remaining.

▶ **Near-Earth asteroid**

The near-Earth asteroid 1999 RQ36 is 1,500 feet (450 m) wide. It has been identified as a potentially dangerous asteroid with a one in 1,000 chance of colliding with Earth in the year 2182. In 2016, NASA plans to launch an unmanned mission to the asteroid to collect samples. This will provide valuable information about its composition and spin, allowing for a more accurate calculation of its likely orbit—and likely future collision with our planet. This artist's conception shows NASA's Osiris-Rex spacecraft approaching the asteroid.

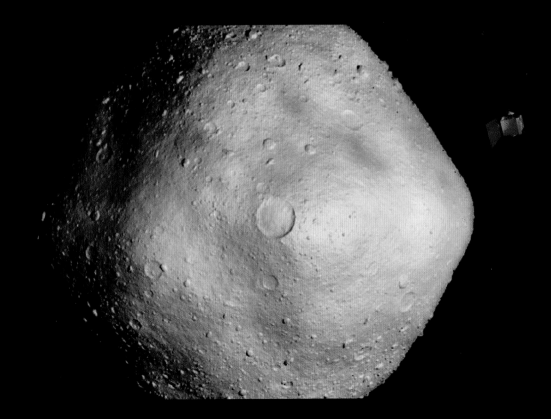

▶ **Quaoar**

This artist's conception shows the large Kuiper Belt object Quaoar. Discovered in 2002, it lies about a billion miles beyond Pluto and completes an almost circular orbit around the Sun every 288 years. Quaoar is 800 miles (1,300 km) in diameter and is a candidate for "dwarf planet" status. It is largely rocky, with a small amount of ice, and is the densest known object in the Kuiper Belt.

▶ **Small objects**

The Hubble telescope has found objects in the Kuiper Belt that are too small to be seen directly. This is an artist's conception of an object with a diameter of about 0.5 miles (1 km). It was detected in 2009 when it passed in front of a background star, temporarily disrupting the starlight. It is the smallest known object in the Kuiper Belt, and was observed at a distance of 4.2 billion miles (6.8 billion km) from Earth.

New discoveries

Asteroids are given numbers when their orbits have been accurately measured. Many asteroids have unusual orbits. The orbit of 433 Eros, which was discovered in 1898, crosses that of Mars. Asteroid 588 Achilles was discovered in 1906 and is locked in a 1:1 resonance with Jupiter, traveling around the Sun in a Jupiter-like orbit that stays about 60 degrees in front of that planet. Asteroid 944 Hidalgo, found in 1920, has an orbit that takes it beyond the orbit of Saturn.

Many of the craters on Mercury, Venus, Earth, the Moon, Mars, and the outer solid planetary satellites have been formed by impacting asteroids. Craters are still being produced, but at a much lower rate than in the past. An asteroid two-thirds of a mile (1 km) wide hits Earth on average every 700,000 years. The direct result would be a crater about 12 miles (20 km) across.

Beyond Neptune

Pluto was downgraded from a planet when it became clear that there was a belt of similar objects beyond Neptune. This means that there are only eight planets in the Solar System. Mercury, Venus, Earth, and Mars are small and rocky. Jupiter, Saturn, Uranus, and Neptune have ice as well, and have captured huge atmospheres of hydrogen and helium. The remainder, including Pluto, the Kuiper Belt, the inner Asteroid Belt, and all the comets, are now believed to be material left over when the planets formed. The material beyond Neptune remains rather mysterious, however. There is still much debate as to whether it was originally formed there, or was thrown out from the Jupiter–Neptune region by gravitational perturbations at some later stage. It is hoped that the arrival of NASA's New Horizons probe in 2015 will answer many of the questions about this mysterious region of the Solar System.

INDEX

Picture Credits

Key: a-above; b-below/bottom; c-center; f-far; l-left; r-right; t-top.

13t and 13b shutterstock.com/Micku, 14b shutterstock.com/ Georgios Kollidas, 16-17b shutterstock.com/Andreas Koeberl, 23b Dreamstime.com/F9 photos, 30-31 Dreamstime.com/Luke Pederson, 52t Dreamstime.com, 130t shutterstock.com/ MarcelClemens, 133 creative commons.

All other images supplied courtesy of NASA.